2017台达杯国际太阳能建筑设计竞赛获奖作品集

Awarded Works from International Solar Building Design Competition 2017

阳光·颐养

SUNSHINE & CARE FOR THE ELDERLY, FOR THE FUTURE

中国可再生能源学会太阳能建筑专业委员会　编

Edited by Special Committee of Solar Buildings, CRES

执行主编：仲继寿　张　磊

Chief Editor: Zhong Jishou, Zhang Lei

编辑：鞠晓磊　夏晶晶　郑晶茹

Editor: Ju Xiaolei, Xia Jingjing, Zheng Jingru

中国建筑工业出版社

CHINA ARCHITECTURE & BUILDING PRESS

图书在版编目(CIP)数据

2017台达杯国际太阳能建筑设计竞赛获奖作品集　阳光·颐养/中国可再生能源学会太阳能建筑专业委员会编.—北京：中国建筑工业出版社，2017.7
ISBN 978-7-112-20980-4

Ⅰ.①2017…　Ⅱ.①中…　Ⅲ.①太阳能住宅-建筑设计-作品集-中国-现代　Ⅳ.①TU241.91

中国版本图书馆CIP数据核字（2017）第152142号

银龄化背景下，在何种建筑中，以何种方式安度晚年为大众所关心。2017台达杯国际太阳能建筑设计竞赛以"阳光·颐养"为主题，分别选取陕西西安和福建泉州赛题，针对生态颐养服务中心进行设计，面向全球征集作品，希望通过优化建筑设计手段，整合适宜的可再生能源技术，让老人沐浴幸福的阳光，让建筑为老人创造安全、健康、舒适、便利、绿色的新生活。

本书可供高等学校建筑设计相关专业本科生、研究生及建筑师参考阅读。

责任编辑：吴　绫　唐　旭　李东禧
责任校对：焦　乐　姜小莲

2017台达杯国际太阳能建筑设计竞赛获奖作品集
阳光·颐养
中国可再生能源学会太阳能建筑专业委员会　编
执行主编：仲继寿　张　磊
编辑：鞠晓磊　夏晶晶　郑晶茹

*

中国建筑工业出版社出版、发行（北京海淀三里河路9号）
各地新华书店、建筑书店经销
北京嘉泰利德公司制版
北京中科印刷有限公司印刷

*

开本：787×1092毫米　1/12　印张：23$\frac{2}{3}$　字数：544千字
2017年8月第一版　2017年8月第一次印刷
定价：178.00元（含光盘）
ISBN 978-7-112-20980-4
（30617）

版权所有　翻印必究
如有印装质量问题，可寄本社退换
（邮政编码 100037）

"老有所养、老有所依、老有所乐"是国人对幸福生活的追求,老龄服务是对中华民族孝、亲、敬、养等传统文化的传承、转化和更新。银龄化背景下,在何种建筑中,以何种方式安度晚年更为大众所关心。本次竞赛以生态颐养服务中心设计为契机,希望通过优化建筑设计手段,整合适宜的可再生能源技术,让老人沐浴幸福的阳光,让建筑为老人创造安全、健康、舒适、便利、绿色的新生活。

感谢台达环境与教育基金会资助举办2017台达杯国际太阳能建筑设计竞赛。

谨以本书献给致力于颐养产业的设计、建设和践行者。

The concept of "looking after elderly people carefully, providing them with dependence and assistance, and making them enjoy happy life" demonstrates Chinese pursuit of happiness, so the elderly service develops into the inheriting, transformation and renewal of traditional Chinese culture including filial piety, intimate relations, respect and living support. Under the background of aging, everyone is paying attention to what kind of building they will live in, and how to spend their twilight years. Through optimizing architectural design means and integrating appropriate renewable energy technology, solar energy can bring sustainable energy for buildings and the buildings can create a safe, healthy, comfortable, convenient, green new life for the old.

Gratitude is given to the Delta Environment and Education Foundation for hosting the International Solar Building Design Competition 2017.

The book aims to pay tribute to those designers, builders and practitioners who are dedicated to the industry of elderly care.

目 录
CONTENTS

阳光·颐养	Sunshine & Care for the Elderly, For the Future
过程回顾	General Background
2017台达杯国际太阳能建筑设计竞赛评审专家介绍 Introduction of Jury Members of International Solar Building Design Competition 2017	

获奖作品　Prize Awarded Works　　001

综合奖·一等奖　General Prize Awarded · First Prize

荼蘼·院落（西安）	Gloomy·Courtyard (Xi'an)	002
风·巷（泉州）	Wind·Cold Lane (Quanzhou)	006

综合奖·二等奖　General Prize Awarded · Second Prize

会聚（西安）	Gather+ (Xi'an)	012
方宅井间（泉州）	Living in between the Patios (Quanzhou)	018
老厝新生（泉州）	Regeneration of Old Dwellings (Quanzhou)	024
光·栖院（泉州）	Light·Habitat & Yard (Quanzhou)	030

综合奖·三等奖　General Prize Awarded · Third Prize

半院（西安）	Half of Yard (Xi'an)	036

颐行颐养（西安）	Roaming in the Sunlight (Xi'an)	042
土生土长（西安）	Half-underground Community (Xi'an)	048
暮·光（西安）	Elderly in Sunshine (Xi'an)	054
诗意栖居（西安）	Poetic Dwelling (Xi'an)	060
安养享阳（泉州）	Enjoy Sunshine Convalesce (Quanzhou)	064

综合奖·优秀奖 General Prize Awarded · Honorable Mention Prize

颐园——老城墙下的怡养空间（西安）	Home for the Old (Xi'an)	068
藤·廊（西安）	Vengevine Arcade (Xi'an)	074
观山居（西安）	Mountain House (Xi'an)	080
阳光花园 生态颐养服务中心（西安）	Sunshine Garden Ecological Remaining Service Center (Xi'an)	086
光院·居（西安）	Sunshine and Yard Build the House (Xi'an)	092
暖院·悟境（西安）	Solar & Silent Yard (Xi'an)	096
沐光之城（西安）	Via Light (Xi'an)	102
漫步·时·光（西安）	Time·Light (Xi'an)	108
悠悠然居（西安）	Leisurely Living (Xi'an)	112
南山·颐居（西安）	Zhongnan Mountain·Residence (Xi'an)	118
秦岭·光居（西安）	Sunshine Elderly Care Center (Xi'an)	124

沐光·团居（西安）	Sunlight·Cluster (Xi'an)	130
温暖的房子（西安）	Warm House (Xi'an)	136
园居安老（西安）	Pastoral Pension Center (Xi'an)	142
逸院昀寮（泉州）	The Wander over Sunshine (Quanzhou)	148
"院"儿里院外（泉州）	In Yard & Out Yard (Quanzhou)	154
禅净·颐养服务中心（泉州）	Calm·Down Maintenance Service Center (Quanzhou)	158
记忆·寮院（泉州）	Sunshine Yard of Memory (Quanzhou)	162
井·巷（泉州）	Well & Lane (Quanzhou)	168
光·转·折（泉州）	Light·Turn·Fold (Quanzhou)	172
围院——太阳能养老院设计（泉州）	Espace Elastique (Quanzhou)	178
游廊串"绿"（泉州）	Play in the Green Corridor (Quanzhou)	184
面朝阳光，随"季"应变（泉州）	Integration Adjustable Bipolar Surface (Quanzhou)	190
光之寓（泉州）	Sanatorium of Sunshine (Quanzhou)	196
园宅院——伍有宅（泉州）	Garden, House, Yard-Five (Quanzhou)	202
墙之庭院（泉州）	The Garden of Walls (Quanzhou)	208
卷光帘（泉州）	Sunshine·Production·Golden-ager (Quanzhou)	214
颐养苑——泉州生态颐养服务中心（泉州）	The Blissful Pure Land (Quanzhou)	220
暮邻·乐居（泉州）	Live with Happiness in Old Age (Quanzhou)	226
光·盒（泉州）	Light Box (Quanzhou)	232

有效作品参赛团队名单	
Name List of all Participants Submitting Effective Works	236
2017台达杯国际太阳能建筑设计竞赛办法	
Competition Brief for International Solar Building Design Competition 2017	248

阳光·颐养
Sunshine & Care for the Elderly, For the Future

当今，我国人口老龄化的趋势日益加快，已经成为世界上老年人口总量最多的国家，未来20年是我国老年人口增长最快的时期，养老服务需求日益提升。我国已初步形成以居家为基础、社区为依托、机构为支撑的养老服务体系，但养老设施数量、环境、结构等方面都尚有不足。不断增长的养老服务需求与相对缓慢的养老服务业发展引起社会各界的高度关注。

西安生态颐养服务中心项目地表景观
Earth Landscape of the Ecological Elderly Care Service Center in Xi'an City Project

国际太阳能建筑竞赛关注当下最具现实意义的热点问题，聚焦"养老服务产业"这一时代的命题，以"阳光·颐养"为主题，组委会分别选取陕西西安和福建泉州赛题，针对生态颐养服务中心面向全球征集作品，希望通过优化建筑设计手段，整合适宜的可再生能源技术，让太阳能为建筑带来永续的能源的同时，为老年人创造安全、健康、舒适、便利、绿色的新生活。两个赛题均定位为养老设施，分属不同气候区，对于竞赛成果在不同地区落地具有较强的现实意义。西安生态颐养服务中心项目结合生态田园养老社区的建设需求，建设适用于寒冷地区的生态颐养服务中心。泉州生态颐养服务中心项目结合德化瓷都印象生态园的定位，建设适用于中亚热带气候区的生态颐养服务中心。

本届竞赛收集到的作品质量较之往届有很大的提升，两个一等奖作品结合赛题，充分考虑老年人的需求，并用与当地气候特点相匹配的主被动技术，具有极高的建筑可实施性和技术经济性。为更好地总结太阳能建筑的设计、教学与实践方法，竞赛组委会首次增设了优秀设计方法奖，鼓励参赛团队从设计方法学角度不断提升，并根据赛题的实际情况设计出最契合的作品。

"梦想照进现实"，获奖作品实地建设是竞赛的一大亮点。在社会各界的支持下，往届竞赛的优胜作品通过深化设计后得以实际建设，其中包括杨家镇台达阳光小学、龙门乡台达阳光初级中学、吴江中达低碳示范住宅和青海农牧民定居农宅等。值得一提的是，自2015届竞赛起，竞赛获奖作品首次实现了成组示范建设。目前，项目的一、二、三等奖和优秀奖获奖作品即将在青海省湟源县兔尔干村建设完成，组委会将组织力量对这些建成项目进行运行测试，为青海乃至全国的新农村建设提供技术支撑。

经过十余年的发展和完善，国际太阳能建筑设计竞赛已经步入到一个新的发展期。竞赛作为行业智慧共享、新能源应用服务、获奖作品实践、创新人才培养和低碳理念传播的综合平台的定位也不断地深入推进。新的时期，竞赛将承载绿色希望，肩负绿色使命，成就绿色梦想。

Currently, the increasingly rising trend of aging population enables China to become the country with the largest population in old age. The next two decades will witness rapid increase of China's aging population and more demands for elderly care service. China has preliminarily developed into the residence-based elderly care service system supported by communities and institutions. However, there are still some shortages in such aspects as the amount of elderly care service facilities, environment, structure and other facilities. As a result, both the increasing demands for elderly care service and relatively slow development of elderly care service industry arouse intensive attention in various fields of the society.

International solar building competitions focus on current hot issues with the most practical significance and the theme of the era of "elderly care service industry". Themed on "Sunshine and Care for the Elderly", the Organizing

Committee of International Solar Building Design Competition adopts contest themes of Xi'an, Shaanxi Province, and Quanzhou, Fujian Province, and collects works worldwide with respect to ecological elderly care service centers. The move, through optimizing architectural design means and integrating appropriate renewable energy technology, aims to create a safe, healthy, comfortable, convenient, green new life for the old while ensuring that solar energy is able to bring sustainable energy for buildings. The two contest themes, oriented towards facilities for the elderly, standard for different climate regions, which enjoys substantially practical significance for contest results in different regions. The Ecological Elderly Care Service Center in Xi'an City project spares no efforts to build a service center appropriate in cold regions concerning the demands for building ecological communities for the elderly care. While the Ecological Elderly Care Service Center in Quanzhou City project strives to build a center appropriate in subtropical climatic regions based on the positioning of Dehua Porcelain Impression Ecological Park.

Quality of the works collected in this competition has made remarkable progress than those of previous years, as the two works winning the first prize are highly operable in construction and economical in technology by giving full consideration to the needs of the elderly given the contest themes, and adopt both active and passive technologies matched with local climate. To draw a better conclusion of designing, teaching and practice methods of solar buildings, the Organizing Committee adds the prize for excellent designing methods for the first time, encourages the teams to make constant improvement from the perspective of designing methodology, and designs the best-quality works according to actual conditions of the contest themes.

Actual construction based on award works highlights the competition with the slogan of "putting the dream into practice". Thanks to support from all sectors of the society, those excellent works from previous competitions have been constructed after in-depth design, including the Delta Sunshine Primary School in Yangjia Town, Delta Sunshine Junior High School in Longmen County, Zhongda Low-carbon demonstration residence in Wujiang City, and farm houses for farmers and herdsmen of Qinghai Province. To be mentionable, it was the first time that the award works have been able to be constructed by groups for demonstration since the competition in 2015. At present, the building projects in Tuergan Village, Huangyuan County, Qinghai Province are going to be completed on the basis of the works winning the first prize, second prize, third prize and excellency award, and the Organizing Committee will join efforts to conduct operation tests for those completed projects, providing technical support for construction of new rural areas in Qinghai and even China.

Over a decade of development and improvement sees a new development period for the International Solar Building Design Competition. As a comprehensive platform on sharing industrial wisdom, service of new energy application, practice of award works, new ways of cultivating talents, and spread of low-carbon concept, the competition will adhere to its positioning and step up more efforts to make progress. It will assume the hope, bear the mission, and achieve the results in developing ecological green landscape.

过程回顾
General Background

2017台达杯国际太阳能建筑设计竞赛由国际太阳能学会、中国可再生能源学会、全国高等学校建筑学学科专业指导委员会主办，国家住宅与居住环境工程技术研究中心、中国可再生能源学会太阳能建筑专业委员会承办，中国建筑设计院有限公司协办，台达环境与教育基金会冠名。在社会各界的大力支持下，竞赛组委会于2016年1月成立，先后组织了竞赛启动、媒体宣传、校园巡讲、作品注册与提交、作品评审等一系列活动。这些活动得到了海内外业界人士的积极响应和参与。

一、竞赛筹备

面对已经到来的"银色浪潮"，在何种建筑中、以何种方式安度晚年成为大众的关注点，本届竞赛题目最终确定为"阳光·颐养"，通过组织专家实地考察，最终选取陕西西安和福建泉州两个真实项目，针对生态颐养服务中心进行设计，确定了设计竞赛的场地建设条件，编制了两个赛题的设计任务书。

二、竞赛启动

2016年6月16日，2017台达杯国际太阳能建筑设计竞赛在北京启动。中国房地产业协会副会长兼秘书长冯俊，中国建筑学会理事长、中国建设科技集团

2017台达杯国际太阳能建筑设计竞赛正式启动
The International Solar Building Design Competition 2017 was launched

This competition is sponsored conjointly by the International Solar Energy Society, the Chinese Renewable Energy Society (CRES) and National Supervision Board of Architectural Education (China), organized by the China National Engineering Research Center for Human Settlements and the Special Committee of Solar Buildings, CRES, and co-organized by the China Architectural Design Institute Co., Ltd with the title sponsor of the Delta Environmental & Educational Foundation. With full cooperation of all the relevant organizations, the Organizing Committee for this competition was set up in January of 2016; they then went on to organize competition start up, media campaigns, campus tours, entry registration and submission, preliminary and final evaluations, technical seminars, etc. These activities have received positive responses and active participation from industry experts at home and abroad.

1. Competition Preparation

In the face of "aging tide", the competition finally sets its them as "Sunshine and Care for the Elderly" as people are paying attention to what kind of building they will live in, and how to spend their twilight years. After field trips of experts, it chooses the two actual projects in Xi'an of Shaanxi Province, and Quanzhou of Fujian Province with respect to ecological elderly care service centers, specifies site construction conditions for designing the competition, and produces task books for designing of the two contest themes.

2. Competition Start-up

On June 16, 2016, the International Solar Building Design Competition 2017 was initiated in Beijing. Many guests attended the opening ceremony of the competition, including Feng Jun, Vice President and Secretary General of the China Real Estate Association, Xiu Long, President of the Council of the Architectural Society of China and Chairman of the Board of the China Construction Technology Consulting Co., Ltd, Wu Qiufeng, Inspector of the China National Working Committee Office on Aging and the former Director of

股份有限公司董事长修龙，全国老龄办巡视员、原事业发展部主任吴秋风，中国可再生能源学会秘书长李宝山，中国可再生能源学会副理事长孟宪淦，台达环境与教育基金会董事长郑崇华，世界银行"中国城市建筑节能和可再生能源应用项目"执行主任田永英，中国可再生能源学会太阳能建筑专业委员会主任委员仲继寿等嘉宾出席并参加了竞赛启动仪式，共同为"2017台达杯国际太阳能建筑设计竞赛"启动揭幕。

本届竞赛分别选取不同气候区，针对生态颐养服务中心面向全球征集作品。其中，西安生态颐养服务中心项目定位于养老设施，为西安及周边地区的健康和轻度失能老年人提供长期养老养生服务。竞赛题目结合生态田园养老社区的建设需求，充分应用太阳能等可再生能源技术，利用周边优越的自然环境，建设适用于寒冷地区的绿色、低碳、健康的生态颐养服务中心；泉州生态颐养服务中心项目定位于养老设施，为福建及周边地区健康和轻度失能老年人提供长期养老养生服务，并向居住在瓷都印象生态园内的老年人提供社区养老服务。本项目结合德化瓷都印象生态园的定位，充分应用太阳能等可再生能源技术，利用周边优越的自然环境，建设适用于中亚热带气候区的绿色、低碳、健康的生态颐养服务中心。

三、校园巡讲

国际太阳能建筑设计竞赛巡讲是本项活动的重要组成部分，自启动以来，得到了清华大学、天津大学、东南大学、重庆大学、山东建筑大学等国内众多建筑院校的大力支持，逐渐成为一项具有影响力的校园公益活动，也吸引了大批富有激情与梦想的青年设计师积极参与竞赛。

2016年9月11日，国际太阳能建筑设计竞赛组委会走进西安科技大学，与师生们围绕国际太阳能建筑设计竞赛进行了交流。竞赛巡讲团一行还前往了西北工业大学、西安交通大学、长安大学和西安建筑科技大学等院校进行校园巡讲。随后，竞赛巡讲团继续前往广西、福建等地，分别在广西大学、桂林理工大学、福州大学、华侨大学、厦门大学开展校园巡讲。巡讲主讲人为国家住宅与居住环境工程技术研究中心曾雁总建筑师，巡讲内容涵盖了太阳能建筑技术应用趋势和现状、历届竞赛获奖作品分析和本届竞赛介绍。通过巡讲，师生们对太阳能建筑设计竞赛和节能技术有了更深入的了解，激发了参赛团队的设计灵感，巡讲后注册人数明显上升。

the Business Development Department, Secretary General Li Baoshan and Vice President of the Council Meng Xiangan of the CRES, Zheng Chonghua, Chairman of the board of the Delta Environmental & Educational Foundation, Tian Yongying, Executive Director of the "Projects on Building Energy Saving and Renewable Energy Application in China's Cities" of the Word Bank, and Zhong Jishou, Chairman of the Special Committee of Solar Buildings, CRES, and they launched the inauguration ceremony for the competition.

The competition adopts different climate regions respectively to collect works worldwide specific to ecological elderly care service centers. The positioning of the Ecological Elderly Care Service Center in Xi'an City project is located on facilities for the elderly. It provides long-term care and health services for people's health and the old with mild disability in Xi'an and surrounding areas. The title of the contest considers the construction demands of ecological rural care community, making full use of solar energy and other renewable energy technologies and surrounding superior natural environment to build a green, low carbon, and healthy ecological maintenance service center that is suitable for the cold regions. The positioning of the Ecological Elderly Care Service Center in Quanzhou City project is also located on facilities for the elderly, which can provide long-term health cultivation service for old people who are health or mild disability in Fujian and the surrounding area and provide community service for the old people living in the Dehua Porcelain Impression Ecological Park. This project is combined with the position of the ecological park, making full use of solar energy and other technologies of renewable energy, utilizing peripheral superior natural environment, to construct green, low-carbon and health service centers for ecological elderly care, which is suitable to subtropical climate regions.

3. Campus Tours

Campus Tours for the International Solar Building Design Competition constitute an integral part of this event. Since its inception, Tsinghua University,

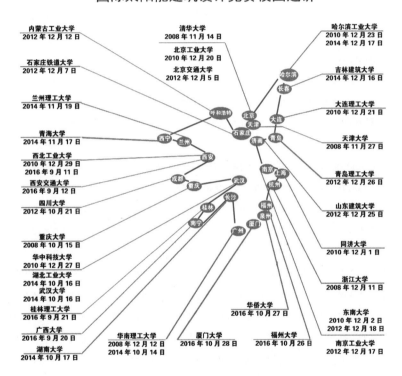

巡讲地图　Map of campus tours

Tianjin University, Southeast University, Chongqing University, Shandong Jianzhu University and many other domestic architectural colleges and universities have given us their support. As a result, the tours have gradually become public benefit activities with much influence on campuses and attracted a large number of passionate young designers to actively participate in the competition.

On September 11, 2016, the Organizing Committee of International Solar Building Design Competition went to Xi'an University of Science and Technology, and made exchanges and communication with teachers and students there centered on the competition. The members of campus tours for the competition also left for several colleges and universities, such as Northwestern Polytechnical University, Xi'an Jiaotong University, Chang'an University, and Xi'an University of Architecture and Technology. Later, they continued to start their trips to some places including Guangxi Province and Fujian Province, and delivered campus tours respectively in Guangxi University, and Guilin University of Technology, as well as Fuzhou University, Huaqiao University, and Xiamen University. Zeng Yan, Chief Architect of the China National Engineering Research Center for Human Settlements, worked as the keynote speaker of campus tours, and made an speech centered on trend and status quo of applying solar building technologies, analysis of award works from previous competitions, and introduction to the competition of this year. The campus tours in which teachers and students enjoyed in-depth understanding of solar building design competitions and energy saving technologies, stimulated aspirations of designing for participant teams, and increasingly more people registered the competition later.

四、媒体宣传

自竞赛启动伊始，组委会通过多渠道开展媒体宣传工作，包括：竞赛双语网站实时报道竞赛进展情况并开展太阳能建筑的科普宣传；在百度设置关键字搜索，方便大众查询，从而更快捷地登陆竞赛网站。在中国《建筑学报》、《建筑技艺》等专业杂志刊登了竞赛活动宣传专版；在《科技日报》、《中国建设报》等30余家平面媒体上发布了竞赛的组织与宣传情况；在新华网、腾讯网、ABBS 等50余家网站上报道或链接了竞赛的相关信息；同时，组委会与诺丁汉大学、谢菲尔德大学等30余所国外院校取得联系并发布了竞赛信息。

4. Media Campaign

Since initiation of the competition, the Organizing Committee has been making efforts to popularize its media influence through many channels, including a bilingual website that reported real-time competition progress

华侨大学巡讲现场
Scenes of campus tour in Huaqiao University

桂林理工大学巡讲现场
Scenes of campus tour in Guilin University of Technology

厦门大学巡讲现场
Scenes of campus tour in Xiamen University

福州大学巡讲现场
Scenes of campus tour in Fuzhou University

西安交通大学巡讲现场
Scenes of campus tour in Xi'an Jiaotong University

西安科技大学巡讲现场 Scenes of campus tour in Xi'an University of Science and Technology

五、竞赛注册及提交情况

本次竞赛的注册时间为2016年6月16日至2017年1月1日，共1193个团队通过竞赛官网进行了注册，其中，中国大陆以外的注册团队25个，包括日本、加拿大、英国等国家和中国港澳台地区。截至2017年3月1日，竞赛组委会收到德国、美国、韩国等国家和中国港、澳、台地区提交的参赛作品239份，其中有效作品232份。

and spread scientific knowledge for solar buildings, set keyword searches on Baidu, enabling public searches much more convenient and much easier to log into the competition website, published special edition for advertising the competition in China's *Architectural Journal*, *Architecture Technique* and other professional magazines, published the information about competition's organization and advertisement in more than 30 printed media platforms including the *Science and Technology Daily*, *China Construction News*, delivered the report or set up links of relevant information for the competition on more than 50 websites, such as Xinhua Net, Tencent and ABBS. Meanwhile, the Organizing Committee reached out to more than 30 foreign Universities, including the University of Nottingham and the University of Sheffield, and release information about the competition.

5. Registration and Works Submission

The registration time of the competition ranged from June 16, 2016 to January 1, 2017, and a total of 1,193 teams made registration via the competition website. Among them, there were 25 registered teams outside the mainland of China, including such countries as Japan, Canada and UK, as well as Hong Kong, Macao and Taiwan regions of China. As of March 1, 2017, the Organizing Committee had received 239 participation works from such countries as Germany, US, and South Korea, as well as China's Hong Kong, Macao and Taiwan regions, 232 of which were valid.

6. Preliminary Evaluation

On March 5, 2017, the Organizing Committee would submit all the valid works to the jury of preliminary evaluation. Each expert would review all the works according to the evaluation requirements regulated in the Competition Evaluation Methods, and select 100 works into the next evaluation process by an order of decreasing number of votes for works. After strict review of those experts, the Organizing Committee would deliver the statistics of evaluation

竞赛官方网站和宣传报道　Oiffcial website and media reports

六、作品初评

2017年3月5日，组委会将全部有效作品提交给初评专家组。每位专家根据竞赛办法中规定的评比标准对每一件作品进行评审，按照作品票数由高到低，共有100份作品进入中评。经过竞赛评审专家的严格审查，组委会对所有专家的评审结果进行统计后，按照作品票数由高到低，共58份作品进入终评阶段。

七、作品终评

竞赛终评会于2017年4月18日在北京召开。在终评会上，经专家组讨论，一致推选喜文华教授担任本次终评工作的评审组长。在他的主持下，评审专家组按照简单多数的原则，集体讨论和公正客观地评选作品，通过三轮的投票，共评选出42项获奖作品，其中一等奖2名、二等奖4名、三等奖6名、优秀奖30名。

results from all the experts, and select 58 works for the final evaluation in accordance with the vote orders from high to low.

7. Final Evaluation

The final evaluation conference was conducted in Beijing on April 18, 2017. Through the discussion of expert groups at the conference, Professor Xi Wenhua was unanimously selected to assume the group leader of evaluating the works in this competition. During his presiding over the conference, the evaluation jury made collective discussion and fair evaluation about the works according to the principle of simple majority. Three-round votes saw 42 award works selected, two of which won the first prize, four of which won the second prize, six of which won the third prize, and 30 of which won the excellence award.

终评会现场　Scenes of final evaluation conference

终评专家组合影　Members of final evaluation jury

2017台达杯国际太阳能建筑设计竞赛评审专家介绍
Introduction of Jury Members of International Solar Building Design Competition 2017

评审专家（终评）
Jury Members (Final Evaluation)

杨经文：汉沙杨有限公司（马来西亚）总裁
Kenneth King Mun YEANG, President of T. R. Hamzah & Yeang Sdn. Bhd.

Ludwig Rongen：德国埃尔福特应用科技大学教授
Ludwig Rongen, Professor of University of Applied Sciences, FH Erfurt

M.Norbert Fisch：德国不伦瑞克理工大学教授，建筑与太阳能技术学院院长
M.Norbert Fisch, Professor of TU Braunschweig and President of the Institute of Architecture and Solar Energy Technology, Germany

林宪德：台湾绿色建筑委员会主席、台湾成功大学建筑系教授
Lin Xiande, Professor of Cheng Kung University, Taiwan

庄惟敏：清华大学建筑学院院长
Zhuang Weimin, Dean of School of Architecture, Tsinghua University

仲继寿：中国可再生能源学会太阳能建筑专业委员会主任委员，国家住宅与居住环境工程技术研究中心主任
Zhong Jishou, Chief Commissioner of Special Committee of Solar Building, CRES and Director of CNERCHS

喜文华：甘肃自然能源研究所所长，联合国工业发展组织国际太阳能技术促进转让中心主任，联合国可再生能源国际专家，国际协调员
Xi Wenhua, Director-General of Gansu Natural Energy Research Institute

冯雅：中国建筑西南设计研究院副总工程师，中国建筑学会建筑热工与节能专业委员会副主任
Feng Ya, Deputy Chief Engineer of Southwest Architecture Design and Research Institute of China

黄秋平：华东建筑设计研究院副总建筑师
Huang Qiuping, Vice-Chief Architect of East China Architecture Design & Research Institute

获奖作品

Prize Awarded Works

综合奖·一等奖
General Prize Awarded · First Prize

注　册　号：4616

项目名称：荼蘼·院落（西安）
　　　　　Gloomy·Courtyard（Xi'an）

作　　　者：季思雨、文瑞琳、王怡璇、
　　　　　张艺冰、王锋宇、张瑞林

参赛单位：河北工业大学
　　　　　石家庄铁道大学
　　　　　西安建筑科技大学
　　　　　中铁建安工程设计院

指导老师：高力强、张　军

专家点评：

作品的建筑平面由三个半开敞且相互联系的庭院组成，公共空间和建筑功能完整，交流流线合理，适宜老年人生活并可满足其医疗需求，建筑造型富有创新和情趣。被动通风、集热、遮阳等技术应用与建筑结合较好。主动技术利用合理，设备转换、传输效率高，建筑可实施性强，技术经济性好。

The building plane is composed of three half-opened and interrelated courtyards, enjoys integral public space and building function, as well as reasonable streamline exchanges, which is appropriate for the elderly life and their demands for medical service. The building shape embodies the features of creativity and charm. Such technologies as passive ventilation, heat collection and sun-shading are applied well into the building. The work utilizes rational active technologies, boasts high efficiency in transmission, sound feasibility in construction, and outstanding technical economy.

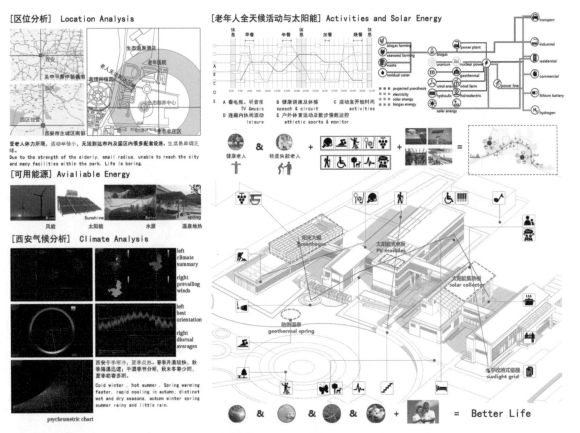

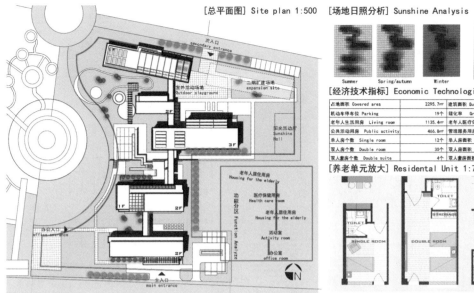

[总平面图] Site plan 1:500

[场地日照分析] Sunshine Analysis

[窗下种植] Planting

荼蘼·院落
gloomy yard 2

[经济技术指标] Economic Technological Index

占地面积 Covered area	2295.7㎡	建筑面积 Building area	3176.6㎡
机动车停车位 Parking	19个	绿化率 Greening rate	56.8%
老年人生活用房 Living room	1135.4㎡	老年人医疗保健用房 Health care	165.4㎡
公共活动用房 Public activity	466.8㎡	管理服务用房 Management service	547.7㎡
单人房个数 Single room	12个	单人房面积 Single room area	14.2㎡/床
双人房个数 Double room	30个	双人房面积 Double room area	28.4㎡/间
双人套房个数 Double suite	4个	双人套房面积 Double suite area	28.0㎡/间

[养老单元放大] Residental Unit 1:75

[活动室部分节能技术分解]

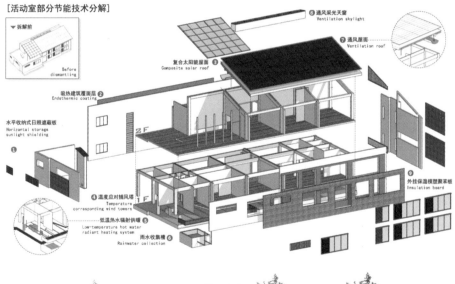

① 水平收纳式日照遮蔽板 Horizontal storage sunlight shielding
② 吸热建筑覆面层 Endothermic coating
③ 复合太阳能屋面 Composite solar roof
④ 温度应对捕风塔 Temperature corresponding wind towers
⑤ 低温热水辐射供暖 Low-temperature hot water radiant heating system
⑥ 雨水收集槽 Rainwater collection
⑦ 通风屋面 Ventilation roof
⑧ 通风采光天窗 Ventilation skylight
⑨ 外挂保温横塑聚苯板 Insulation board

[技术详解] Technics

复合太阳能屋面结构 Composite solar roof
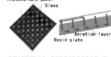

利用光伏板输出电力，集热板及通气层吸收热量，同时透过玻璃采光将日光导入建筑的目标。
The utility model uses solar energy to provide power and heat collecting plate and the ventilation layer to make the sunlight is introduced into the building through the glass.

温度应对捕风塔 Temperature corresponding wind towers

根据温度不同调整捕风塔的通风效果。
According to the different temperature, adjust the wind tower ventilation effect.

光导照明系统 Photoconductive lighting system

By using light-absarting shade and efficient light transmission system, it can provide good illumination for the space where is lacks of delighting, therafore making the interior lighting environment more natural and comfortable.

通风楼板 Ventilation floor

As a part of the whole building ventilation system, the ventilation folir can easily make the waste air changed by the dry and fresh air, bringing comfortable enviroment with low carbon cost.

雨水收集装置 Rainwater collection

既可以有效收集雨水又可以合理节约成本，可以用到生活中的杂用水。节约自来水，减少水处理的成本。
Not only can effectively collect rainwater and reasonable cost savings. Can be used in life miscellaneous water, save water, reduce the cost of water treatment.

[南立面图] South Elevation 1:300

[东立面图] East Elevation 1:300

[西立面图] West Elevation 1:300

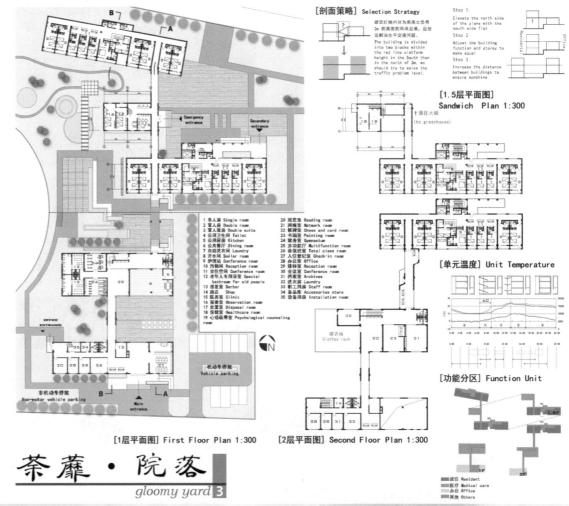

荼蘼·院落
gloomy yard 3

千顷菜畦十程洲
溪居宜月更宜秋
鸥鹭栖水高僧舍
鹃鸪紫云名士楼
苍菊紫芬飞鹭羽
获雀花敷钓鱼舟
黄橙红柿紫菱角
不羡人间万户侯

[水平收纳式日照遮蔽板] Sunshine Shield

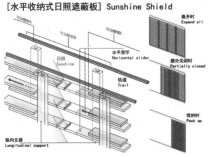

利用一条能够在水平方向延伸的夹具，与阳光的屏蔽能力作出小调整。在水平方向上滑动提高遮蔽强度，不易被风影响。此外，在夜间起防范作用，并可以保持窗口打开。

With the shielding ability of sunshine to make small adjustment. It is the height of each slat and as fixture capable of extending in the horizontal direction. Store in the horizontal direction, the slat is possible to have a high strength, difficult to be fanned by the wind. In addition, the role of intrusion prevention and in the night, and can keep the window open.

[剖面策略] Section Strategy

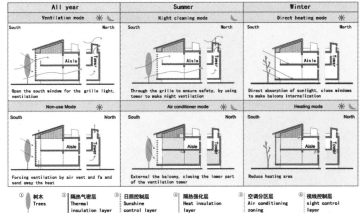

通过六个层次的改变，实现建筑对环境的应答，达到室内环境符合人的需求的目标。各个层次不同策略的排列组合，极大地提高了建筑的可变性，改善了舒适度。

Through the six levels of change, to achieve the construction of the response to the environment, to meet the needs of people in the building environment. The arrangement and combination of different strategies at different levels greatly improve the variability of the building and improve the comfort.

[单元内能源系统] Unit Energy System

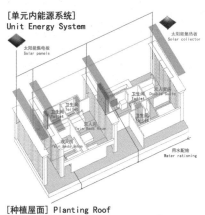

[日影分析] Shadow Map

Summer solstice　Winter solstice　Equinoxes

[日照模拟] Sunlight Simulation

 Unit south elevation

 Health care centre elevation

 Health care centre elevation

荼蘼·院落 gloomy yard 4

[种植屋面] Planting Roof

种植屋顶以更便宜的方式使内部冷却和温室取暖，并同时产生收益。
Planting roof makes inside cooler and greenhouse warmer by a cheaper way. It also produce gains.

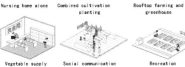

通过屋顶收集热量，并以墙体内的空气作为介质将热量传导至地板下实现供暖，减少热量的浪费。
The heat is collected through the roof, and the air in the wall is used as a medium to conduct heat to the floor to realize heating and reduce the waste of heat.

屋顶设置小块农田及温室，以此为契机，起到提供部分蔬菜、提供交流机会、休闲娱乐的作用。
The roof is a small piece of farmland and greenhouse, as an opportunity to play a part of the supply of vegetables, providing opportunities for exchange, leisure and entertainment.

[T形屋顶的探讨] Discussion on T Roof

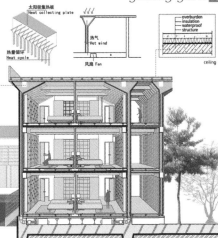

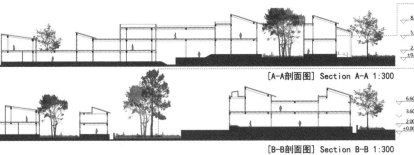

[A-A剖面图] Section A-A 1:300

[B-B剖面图] Section B-B 1:300

[透光蓄热窗] Heat Storage Plate

白天将太阳光及部分热量收集，在夜间为室内供暖。
Use of the heat absorbed by the glass during the night for indoor heating.

[湿地景观] Wetland Landscape

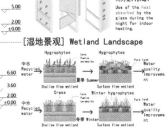

综合奖·一等奖
General Prize Awarded · First Prize

注 册 号：4689
项目名称：风·巷（泉州）
　　　　　Wind · Cold Lane (Quanzhou)
作　　者：高海伦、董兆、梅博涵
参赛单位：浙江理工大学
指导老师：文强

专家点评：

作品总平面布局结合场地特点及任务书要求，出入口、广场、二期发展用地预留等考虑细致、周到。南北功能分区明确，动线简单明了，比较符合老年人生活需求。建筑立面采用当地材料，统一和谐而富于变化。通过建筑构造手段解决采光、通风问题，符合当地气候特点，主被动技术与建筑设计融洽，具有较高的可实施性和经济性。

The overall plane layout, concerning the site characteristics and requirements for the task book, takes into account such land reservation as entrance and exit, square, the second-stage land use for development. The work makes specific in the division of south-north function, and clear in generatrix, relatively satisfying the needs of the elderly life. The building elevation adopts local materials which are harmonious and variable. To solve the problem on lighting and ventilation through building structure complies with local climate features. The active and passive technologies are applied well with the combination with building design, demonstrating relatively high feasibility and economy.

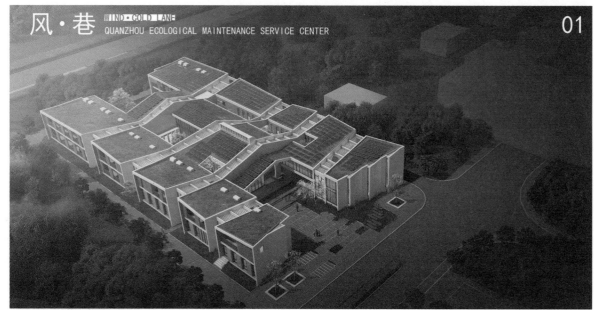

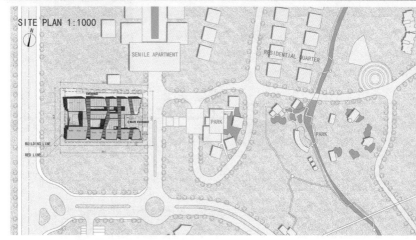

CHAPTER 1 PRE-PROGRAM

1.1 BACKGROUND ANALYSIS

● QUANZHOU IMAGE ANALYSIS

● SITE ANALYSIS

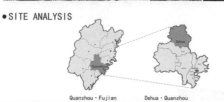

1.2 TRADITIONAL ARCHITECTURAL ANALYSIS

● ENHANCE THE AIR FLOW

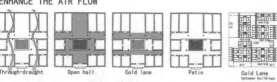

● AVOID DIRECT SUNLIGHT

● ROOF RAINWATER DRAINAGE

风·巷
WIND·COLD LANE
QUANZHOU ECOLOGICAL MAINTENANCE SERVICE CENTER

02

- Quanzhou Traditional Architecture
- Modern Sanatorium
- Breaking the Scale

● SITE PHOTOS

1. Lane area: 4928m²

2. The building is square according to the shape of the site.

CHAPTER 2: CONCEPT

● MANUAL MODEL

3. Function zoning and Patio.

4. Insert the cold lane and split the building volume.
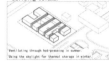

5. The cold lane is deformed to form the "Venturi Effect"

6. Sloping roof is conducive to placing solar panels.

7. Add skylights to improve indoor lighting and ventilation.

8. Further refine the building.

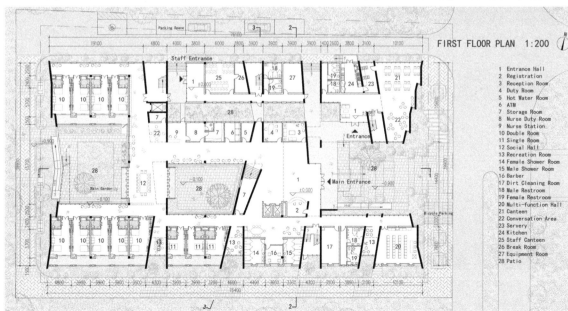

FIRST FLOOR PLAN 1:200

1 Entrance Hall
2 Registration
3 Reception Room
4 Duty Room
5 Hot Water Room
6 ATM
7 Storage Room
8 Nurse Duty Room
9 Nurse Station
10 Double Room
11 Single Room
12 Social Hall
13 Recreation Room
14 Female Shower Room
15 Male Shower Room
16 Barber
17 Dirt Cleaning Room
18 Male Restroom
19 Female Restroom
20 Multi-function Hall
21 Canteen
22 Conversation Area
23 Servery
24 Kitchen
25 Staff Canteen
26 Break Room
27 Equipment Room
28 Patio

风·巷 WIND·COLD LANE
QUANZHOU ECOLOGICAL MAINTENANCE SERVICE CENTER

03

SECOND FLOOR PLAN 1:200

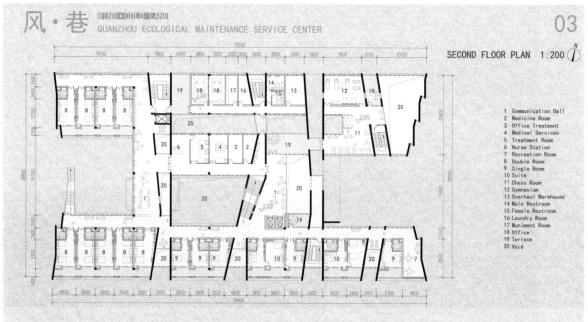

1. Communication Hall
2. Medicine Room
3. Office Treatment
4. Medical Services
5. Treatment Room
6. Nurse Station
7. Recreation Room
8. Double Room
9. Single Room
10. Suite
11. Chess Room
12. Gymnasium
13. Overhaul Warehouse
14. Male Restroom
15. Female Restroom
16. Laundry Room
17. Muniment Room
18. Office
19. Terrace
20. Void

• DOUBLE ROOM PLAN
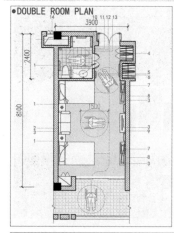

• SINGLE ROOM PLAN

1. Call Button
2. Phonepatch
3. Socket
4. Wardrobe
5. Light Switch
6. Air-Condition
7. Wheelchair
8. Shelf
9. TV Reception
10. Energy Saving
11. Restroom Light
12. Corridor Light
13. Safety Rail

• AXONOMETRIC DRAWING
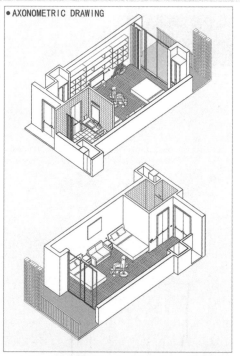

• ELEVATION OF THE ROOM
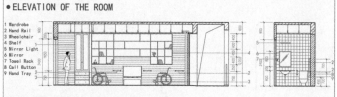

1. Wardrobe
2. Hand Rail
3. Wheelchair
4. Shelf
5. Mirror Light
6. Mirror
7. Towel Rack
8. Call Button
9. Hand Tray

• HUMAN SCALE & UNITS EVERY SQUARE IS 200mm×200mm
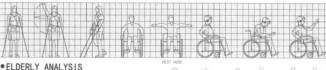

• ELDERLY ANALYSIS

CHAPTER 3: HUMANISM

● 1-1 SECTION 1:100

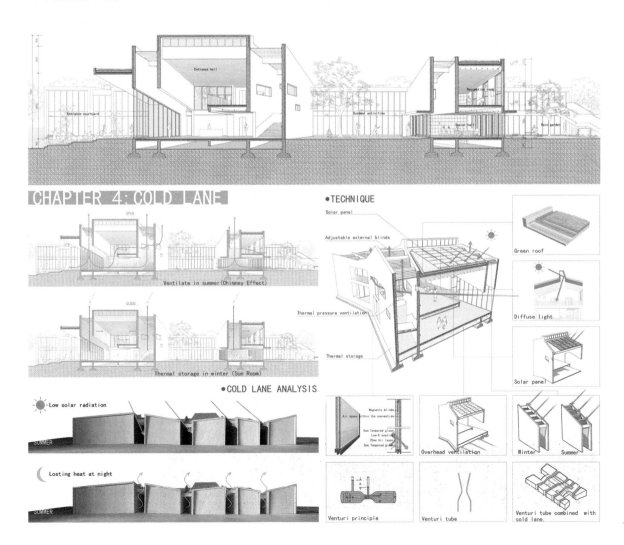

风·巷 WIND·COLD LANE
QUANZHOU ECOLOGICAL MAINTENANCE SERVICE CENTER

● 2-2 SECTION 1:100

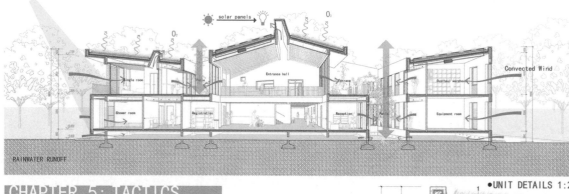

CHAPTER 5: TACTICS

● UNIT DETAILS 1:25

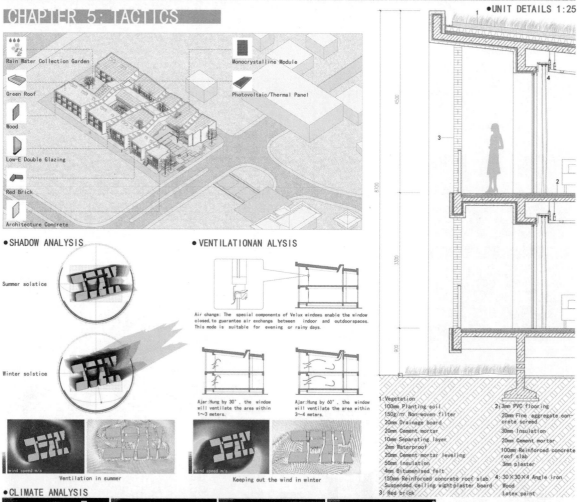

● CLIMATE ANALYSIS

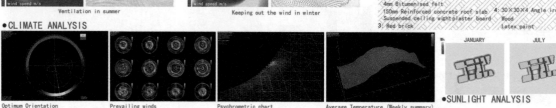

● SUNLIGHT ANALYSIS

风·巷 WIND·COLD LANE
QUANZHOU ECOLOGICAL MAINTENANCE SERVICE CENTER

06　2017 台达杯国际太阳能建筑设计竞赛获奖作品集

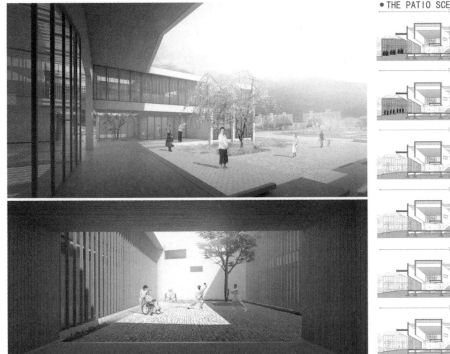

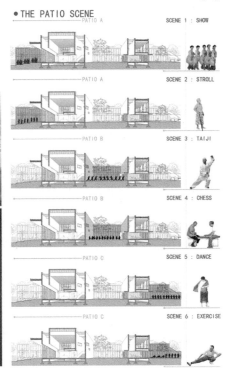

• THE PATIO SCENE

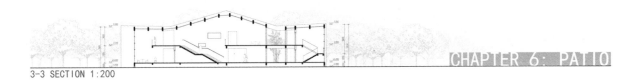
3-3 SECTION 1:200

CHAPTER 6: PATIO

EAST ELEVATION 1:200

SOUTH ELEVATION 1:200

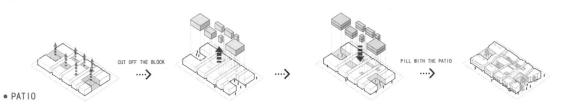
• PATIO — CUT OFF THE BLOCK → FILL WITH THE PATIO →

011

综合奖·二等奖
General Prize Awarded ·
Second Prize

注 册 号：5585
项目名称：会聚（西安）
　　　　　Gather+（Xi'an）
作　 者：陆垠、顾晨
参赛单位：南京工业大学
指导老师：张海燕

专家点评：
作品从中国传统的建筑布局和建筑造型中获得灵感，采用家庭模式的布局，促进邻里交往，采用传统的建筑构造和建筑材料，被动式太阳能技术的应用较为合理。太阳能、地源热泵等主动式技术的应用合理。但作品缺少集中活动和交往空间，未考虑未来的扩建设计。居住空间自然通风欠考虑。

The work inspired by traditional Chinese architectural composition and modeling adopts the large-scale domestic layout to facilitate communication among neighbors, and adopts traditional architectural structure and materials to apply passive solar energy technology rather reasonably. The application of such active technologies as solar energy and ground coupled heat pump is reasonable. However, the work lacks intensive space for activities and exchanges, without considering the extension plan in the future, and placing rather few thoughts on natural ventilation of living space.

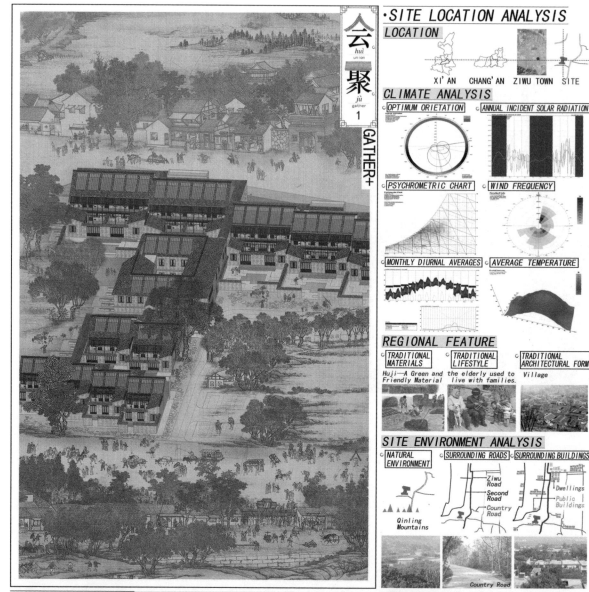

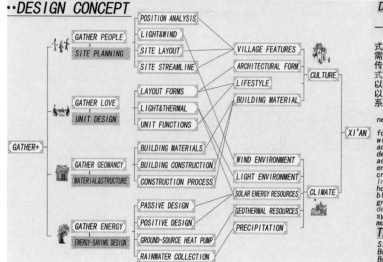

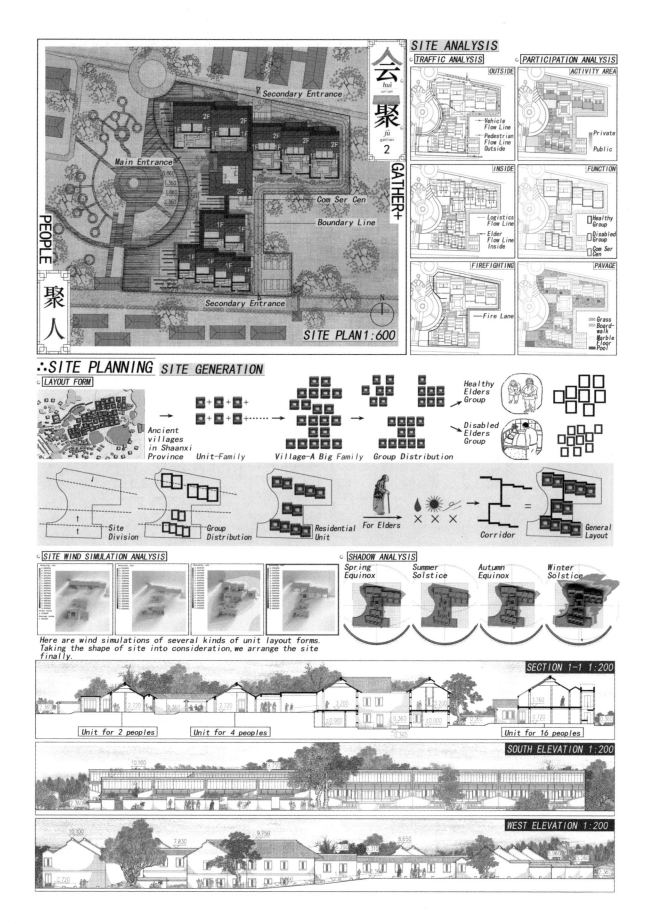

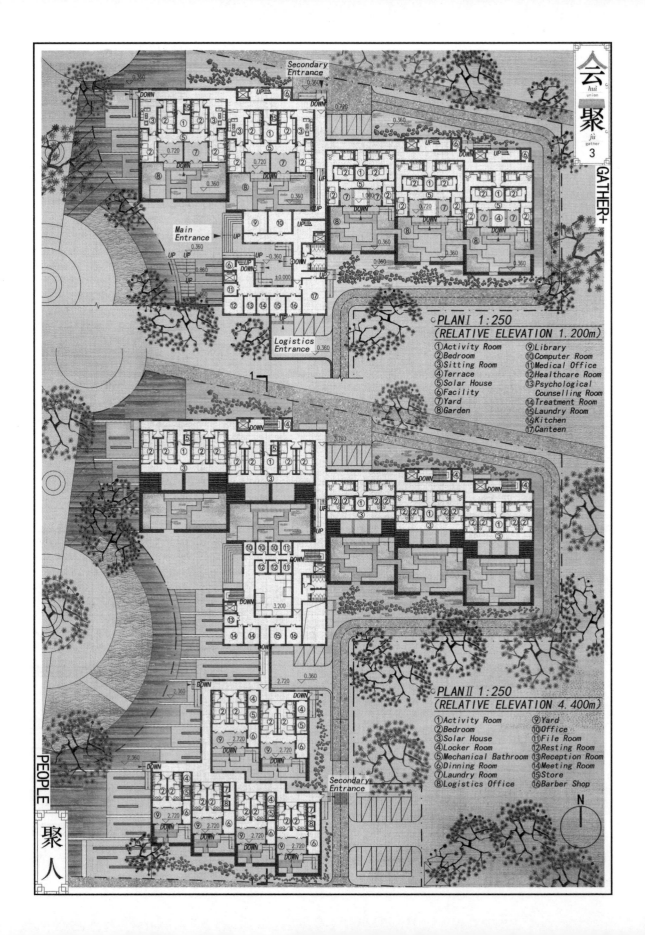

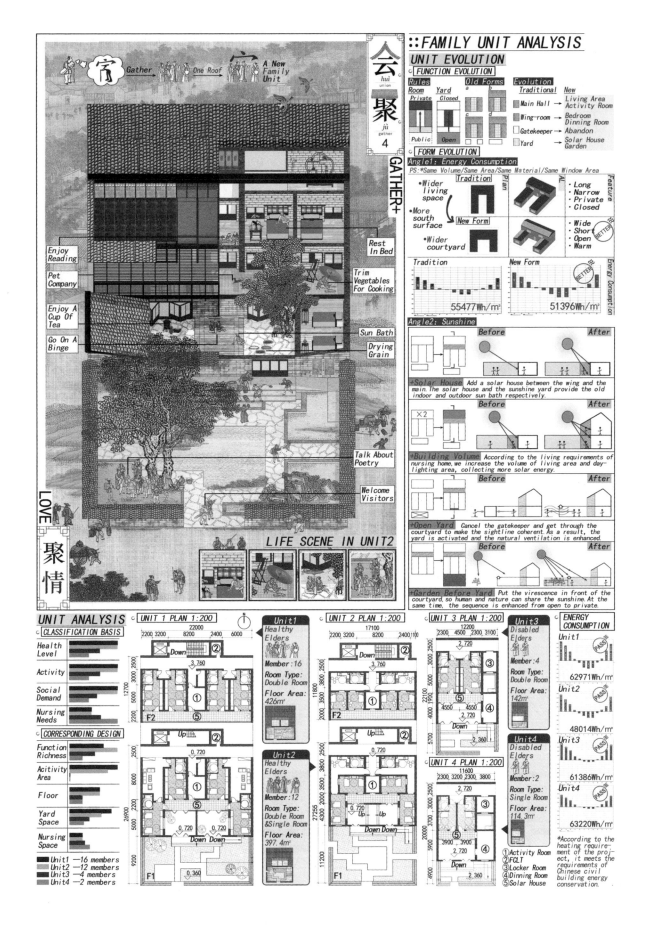

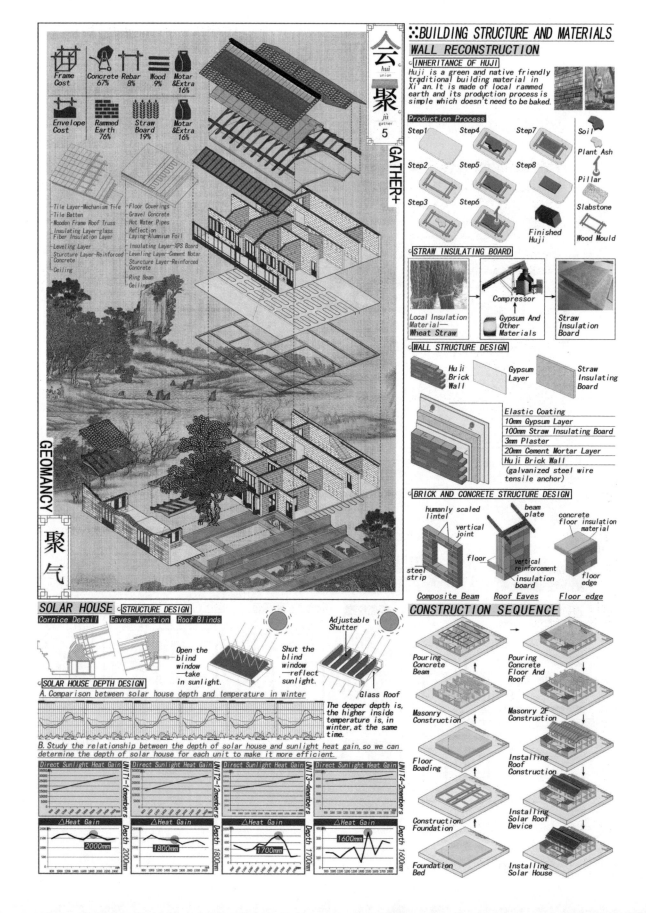

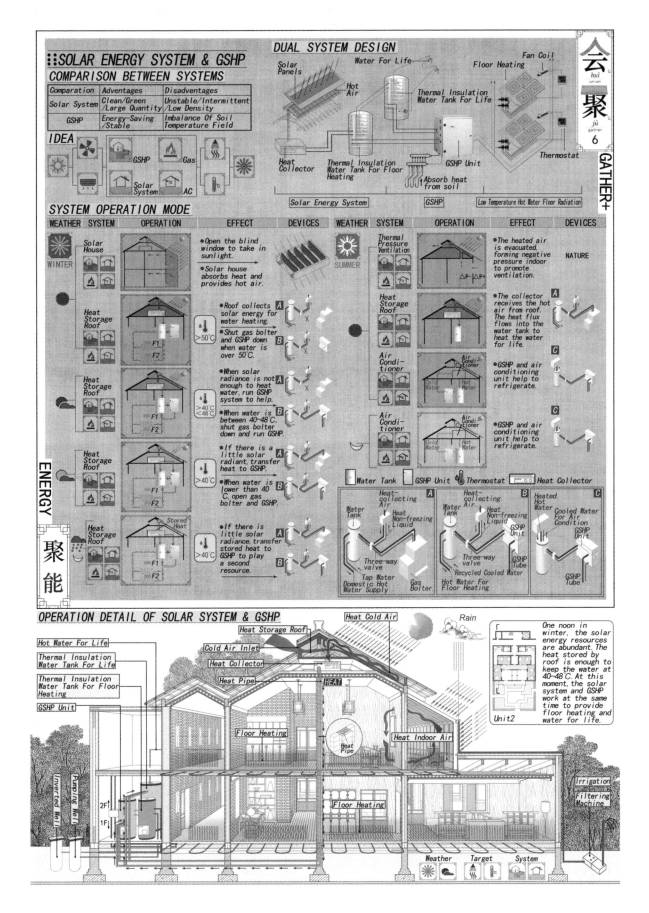

综合奖·二等奖
General Prize Awarded · Second Prize

注册号：4642
项目名称：方宅井间（泉州）
　　　　　Living in between the Patios (Quanzhou)
作　　者：刘程明、王莹莹、叶葭、全孝莉
参赛单位：天津大学
指导老师：刘彤彤、张颀

专家点评：

作品整体建筑基于村庄设计，个人公共空间的紧凑型建筑设计能给人们的生活带来更多便利，庭院设计使人们可以在室外遮阳乘凉，提升舒适感。作品对自然通风、自然光照以及被动、主动太阳能使用方面进行了详细分析，建筑的布局可以保证在公共空间达到高度的自然通风，对于屋面的设计和变化富有特色。

The general architecture is based on a village design. Compact design of the buildings with individual public places creates an attractive living. The courtyard provides self shading and create high comfort in the outdoor environment. The work gives detail analysis about natural ventilation, natural light, and passive/active solar use, so the arrangement of the buildings ensure a high-level of natural ventilation in the public spaces. In this way, the measure makes the roof characteristic in design and changes.

2017台达杯国际太阳能建筑设计竞赛

方宅井间 01
Living in between the Patios

Site Location Analysis

Climate Analysis

方案以12m×12m的单体为设计出发点，结合老年人生活特征、泉州传统民居形式进行建筑设计。重点加强建筑的自然通风处理，以泉州传统民居手巾寮、大厝为原型设计来源，将庭院、天井设计、建筑拔风处理、增加灰空间等手法综合运用。我们尝试加入了雨水收集、中水循环利用的处理手法；将东西朝向的立面、露天平台上增加绿植，利用收集的雨水以及循环的中水来浇灌。为调节室内温度、通风换气，南北向墙体引入特朗伯墙体，改善室内物理环境。

The scheme takes the monomer of 12m×12m as the starting point of the design, combining with the characteristics of the old people's life and the design of the traditional residence(Shoujinliao, Dacuo prototype) in Quanzhou. We try to focus on strengthening the natural ventilation, try to increase the comprehensive use of the yard and patio design, building wind drawing processing, techniques such as gray space and the solar heating curtain walls. We also try to add the rainwater collection and water recycling treatment; the plants on the walls and platforms; rainwater collection and utilization of recycling water.

| High Density & Multi Orientation | Traditional Chinese Curved Roof | Narrow Courtyard | Brick & Stone |

Vertical Type　Square Type　Rectangle Type　Ventilation and Lighting　Drainage　Grey Space

Features of Regional Architectural

Energy—Application of Solar Panels　Lighting—Skylight and Atrium　Ventilation—Wall & Roof & Patio　Water—Rainwater Collection Reclaimed Water Circulation　Plants—The Wall and Roof　Public Space & Unit Classification & Traditional Features

Strategy of Architectural Design

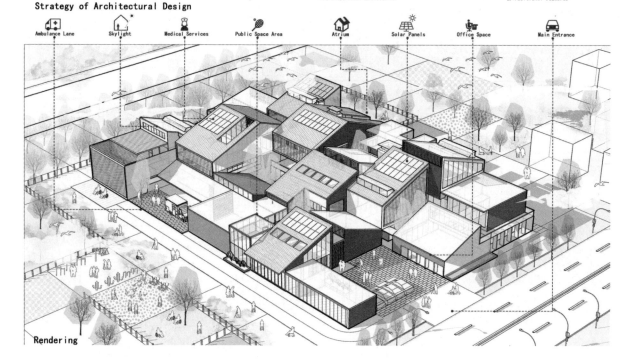

Rendering

2017台达杯国际太阳能建筑设计竞赛

方宅井间02
Living in between the Patios

Three kinds of plane shape and texture map of Fuzhou street seven square of traditional courtyard houses.

Connection arrangement form and building large area within the scope of the patio group

Grey space ratio 54.11%

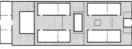

PLAN

And the elderly activity features of traditional architectural features, the design of single prototype will eventually focus on combining the courtyard space of 12 square meters. Specific functional groups are placed in the gray space and location design of the courtyard, placed to light demanding living space according to the prototype, a group arrangement deepening. Many times to refine the plane, adjust the shape of the lighting, ventilation calculation, according to the software testing and specific problems to solve, adjust the program

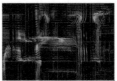
First Plane ventilation (Wind vector)

First Plane ventilation (Wind chart)

Profile 1-1 (Wind chart)

Second Plane ventilation (Wind vector)

Second Plane ventilation (Wind chart)

Profile 2-2 (Wind chart)

Site Plan 1:500

Model

Shou jin liao form

Courtyard combination

Planar mesh

Square combination

Square stacking

Square rotation

Quanzhou patio type "Shoujinliao" building and courtyard building "Da Cuo" is a typical regional architecture, this life mode for local people's behavior and psychology has produced a great influence.
This design comes from Shoujinliao and Dacuo, we use for reference the courtyard space of home building and the space organization. For the elderly, the courtyard space will become the largest outside the bedroom in addition to its pleasant environment.

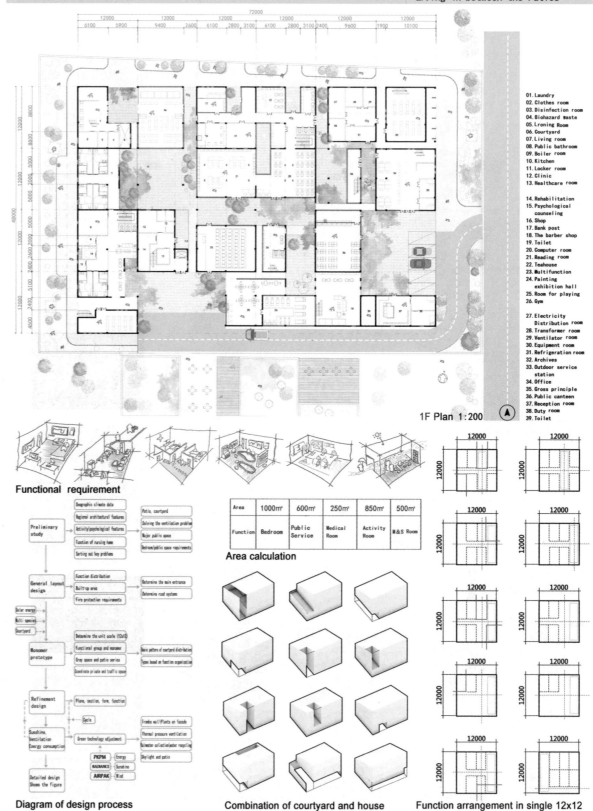

2017台达杯国际太阳能建筑设计竞赛 | 方宅井间04
Living in between the Patios

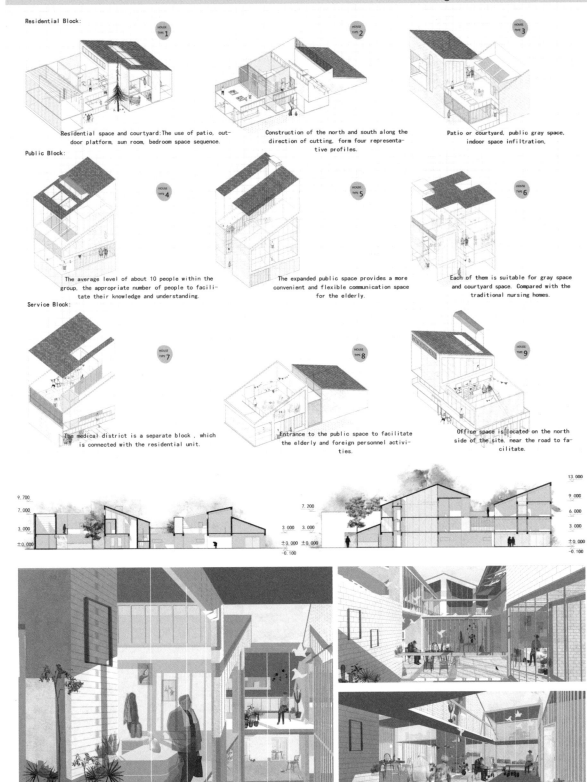

Residential Block:

HOUSE TYPE 1 — Residential space and courtyard: The use of patio, outdoor platform, sun room, bedroom space sequence.

HOUSE TYPE 2 — Construction of the north and south along the direction of cutting, form four representative profiles.

HOUSE TYPE 3 — Patio or courtyard, public gray space, indoor space infiltration.

Public Block:

HOUSE TYPE 4 — The average level of about 10 people within the group, the appropriate number of people to facilitate their knowledge and understanding.

HOUSE TYPE 5 — The expanded public space provides a more convenient and flexible communication space for the elderly.

HOUSE TYPE 6 — Each of them is suitable for gray space and courtyard space. Compared with the traditional nursing homes.

Service Block:

HOUSE TYPE 7 — The medical district is a separate block, which is connected with the residential unit.

HOUSE TYPE 8 — Entrance to the public space to facilitate the elderly and foreign personnel activities.

HOUSE TYPE 9 — Office space is located on the north side of the site, near the road to facilitate.

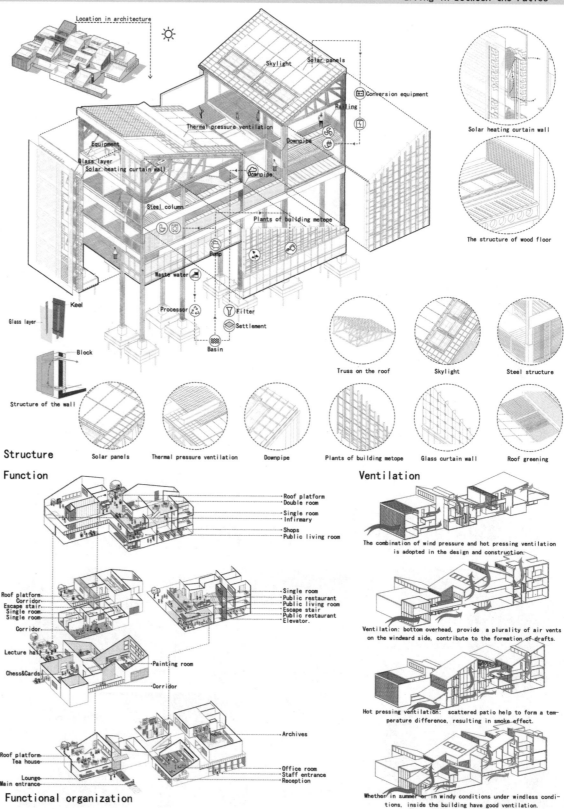

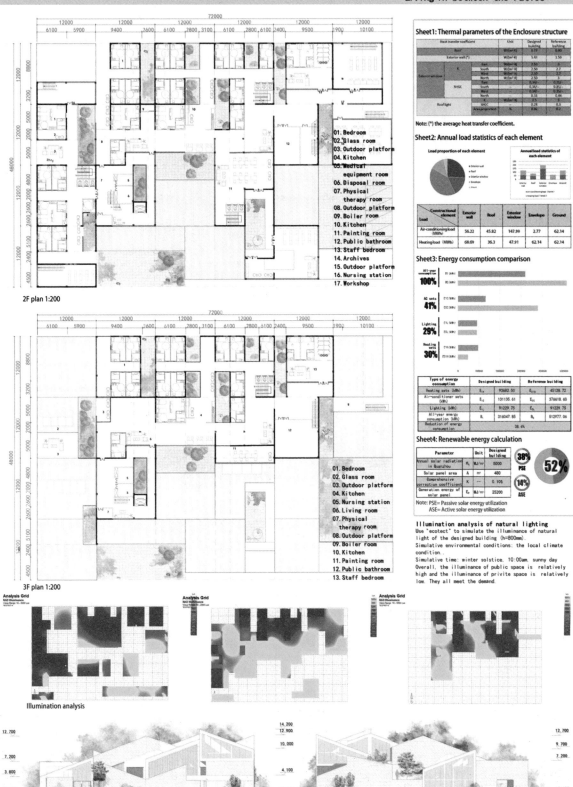

老厝新生 REGENERATION OF OLD DWELLINGS 01

综合奖·二等奖
General Prize Awarded · Second Prize

注 册 号：4896
项 目 名 称：老厝新生（泉州）
　　　　　　Regeneration of Old Dwellings (Quanzhou)
作　　　者：张礼陶、柏陈威、申晓艺
参 赛 单 位：华中科技大学
指 导 老 师：徐　燊

专家点评：

作品保留了传统的建筑风格，建筑连廊促进了交流性，使老人拥有在家居住的感觉，提升居住房间的归属感，利用坡屋面的高度变化实现自然通风、采光和遮阳。入口设计不够明显，走廊交通不够便利，应进一步改善以便于老人出行。提出了地源热泵、太阳能集热器以及沼气系统等能源概念，但系统设计过于复杂，不宜实施。

The building form reminds traditional building style. It encourages social interaction along the corridor. It helps senior person to get home-feeling and identify their own room. And the building presents the landscape of natural ventilation, abundant daylight and sunshading through the variation of height of pitched roofs. However, the entrance is not clearly defined and the corridors are rather inconvenient, so it should be improved to be more welcome to senior person. It is inadvisable for the implementation of the energy concept which seems too complex – ground coupled heat pump, solar thermal collector, and biogas system are proposed.

Field Background

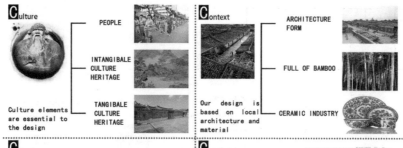

Climate

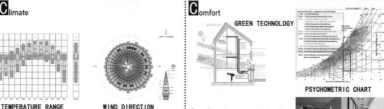

Comfort

设计说明

设计从泉州当地的古厝出发，借鉴其组合方式与"假厝"即双层屋顶的防晒隔热模式。巧妙利用"假厝"南、北向的不同户型，借助房间的组合模式，达到防晒隔热与围合院落空间的双重目的并同时兼顾了夏季通风、冬季抵挡寒风以及防潮与采光等问题。

技术方面，建筑兼顾主动式与被动式的太阳能技术。并采取了地源热泵、雨水花园与沼气系统等其他绿色技术。

材料方面，由于地处瓷都，当地产生很多陶瓷工业的废料，我们将这些废料回收再加工，制成多种多样的建筑材料，应用于建筑之中。

养老设计方面，我们注重老龄化功能安排与空间尺寸，并提供了完备的无障碍配套设施。

Design Instruction

Starting from the design of Quanzhou local ancient houses from their combination with "false Cuo" double roof SPF insulation model. The clever use of "fake Cuo" different apartment layout on South and North, with a combination of model room, achieve the aims of SPF insulation and enclosed courtyard space and at the same time, out of the cold wind in winter and summer ventilation and lighting problems such as moisture. Technical aspects, building both active and passive solar energy technology. And take the ground source heat pump, rain garden and biogas system and other green technology.
Materials, due to a lot of waste porcelain, ceramic industry and local produce, we will be the waste recycling processing, made of various building materials used in construction.
Pension design, we focus on the aging of the functional arrangements and the size of space, and provides a complete barrier free facilities.

Technology Utilization

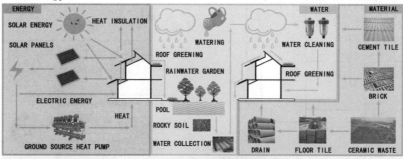

Aerial View

老厝新生 REGENERATION OF OLD DWELLINGS   02

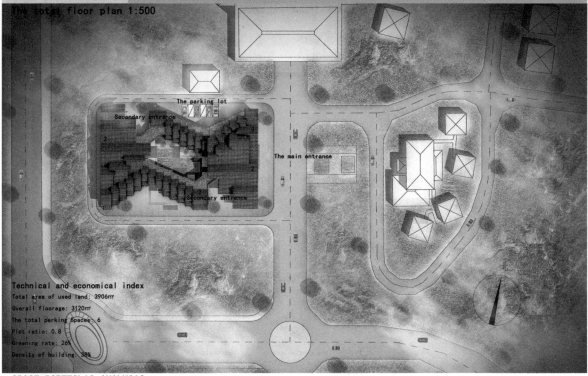

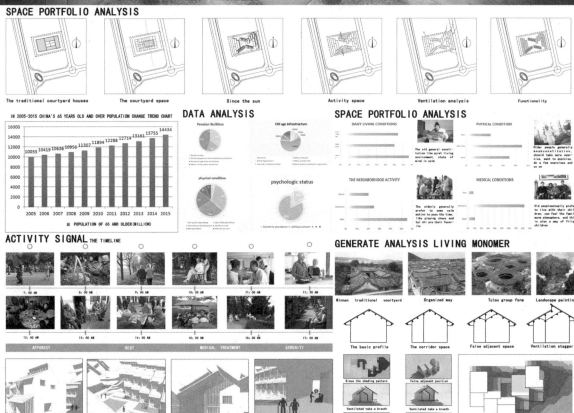

老厝新生 REGENERATION OF OLD DWELLINGS

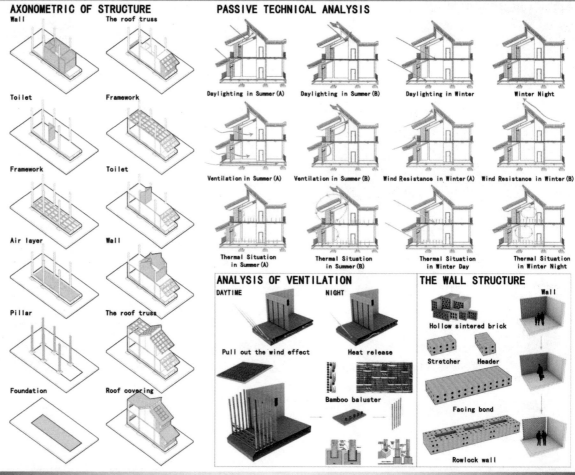

PROFILE CONTROL PERSPECTIVE A

老厉新生 REGENERATION OF OLD DWELLINGS 04

SINGLE ROOM 1:100

DOUBLE ROOM 1:100

DOUBLE SUITE 1:100

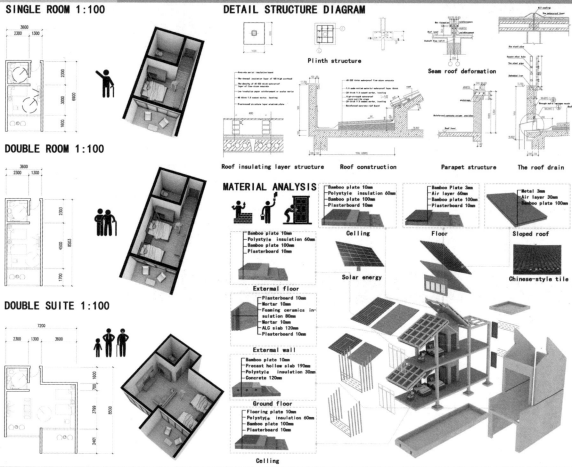

DETAIL STRUCTURE DIAGRAM

MATERIAL ANALYSIS

PROFILE CONTROL PERSPECTIVE B

老厭新生 REGENERATION OF OLD DWELLINGS   05

FIRST FLOOR PLAN 1:200

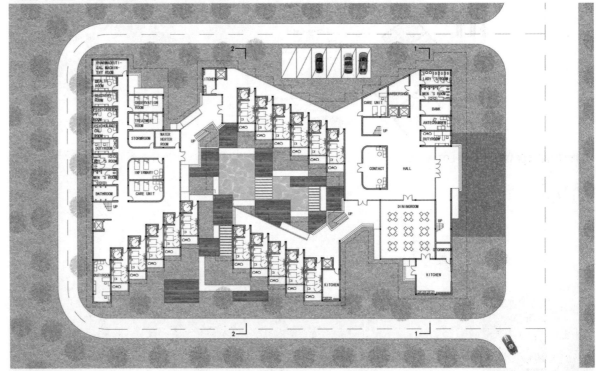

1-1 SECTION 1:200   **2-2 SECTION 1:200**

老厦新生 REGENERATION OF OLD DWELLINGS　06

ENTRANCE EFFECT CHART

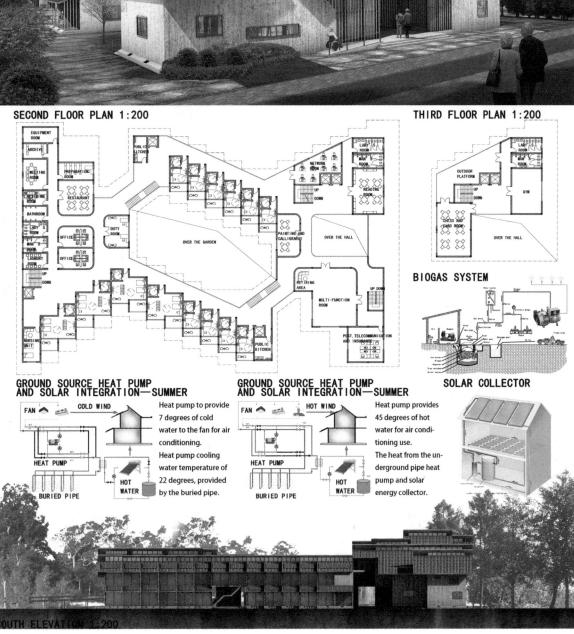

SECOND FLOOR PLAN 1:200

THIRD FLOOR PLAN 1:200

BIOGAS SYSTEM

GROUND SOURCE HEAT PUMP AND SOLAR INTEGRATION—SUMMER

Heat pump to provide 7 degrees of cold water to the fan for air conditioning. Heat pump cooling water temperature of 22 degrees, provided by the buried pipe.

GROUND SOURCE HEAT PUMP AND SOLAR INTEGRATION—SUMMER

Heat pump provides 45 degrees of hot water for air conditioning use. The heat from the underground pipe heat pump and solar energy collector.

SOLAR COLLECTOR

SOUTH ELEVATION 1:200

光·栖院 LIGHT · HABITAT & YARD

综合奖·二等奖
General Prize Awarded · Second Prize

注　册　号：5404
项目名称：光·栖院（泉州）
　　　　　Light·Habitat & Yard
　　　　　（Quanzhou）
作　　者：郎　冰、刘　刚、李思萌
参赛单位：山东建筑大学
指导老师：崔艳秋、薛一冰

专家点评：

作品整体设计紧凑，建筑形体规矩，具有地域建筑特点。作品平面布局公共领域与私人领域分开的做法值得肯定，两个庭院之间能实现自然通风，营造更加舒适的小气候。被动式太阳能技术运营较为合理，在主动式技术方面，提出了地源热泵、太阳能集热器以及光伏系统等概念，但能源系统设计不够清晰，缺乏实用性。

The compact design results in a large-scale building performance. The separation in public and private areas are positive. The two courtyards allow a good circulation and provide comfortable micro-climate. The operation of passive solar energy technology is relatively reasonable. However, concerning active technologies, the energy concept which is unclear lacks practicability because of a mixture of various technologies: ground coupled heat pump, solar thermal collector, and PV system are proposed.

泉州生态颐养服务中心设计 1

Geography Analysis

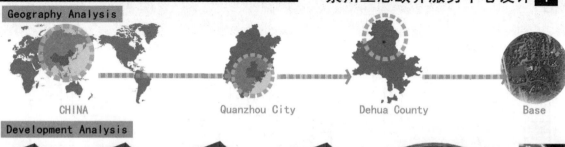

CHINA　　Quanzhou City　　Dehua County　　Base

Development Analysis

    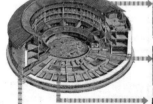

extract traditional element: "Fujian earth dwellings" | exchange the shape to adapt the base | break the shape | move the shape

put the gallery into the room | open to the south to guide the wind and build a place for activities | change the roof | change the roof and building to guide the light come into the room

RAMMED EARTH WALL

TILE
REDWOOD
BAMBOO

Concept Analysis

Inspiration And development

  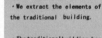

 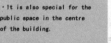

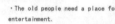

WHAT THEY NEED?

· The old people need more care.

· The old people need a place for entertainment.

· The old people need a comfortable enviroment.

· We extract the elements of the traditional building.

· The traditional building has a simple geometric shape and the space is concentric.

· It is also special for the public space in the centre of the building.

光·栖院 LIGHT · HABITAT & YARD

CLIMATE SIMULATION

The best toward architectural design

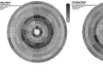

The wind of spring, summer, autumn and winter

Enthalpy wet figure

MODEL OVERVIEW

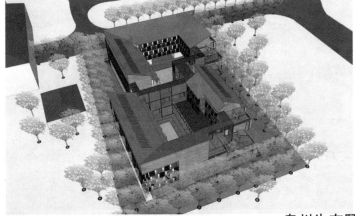

DESIGN INSTRUCTION

The idea comes from the traditional residential building called "tulou". Tulou is in Fujian Province. The yard is the core of the space, it is also the core of the daily life. Our yard is surrounded by two parts and it is open to the outside. The two parts is connected by a porch, people in the yard can see each other through the porch and they can also stay privative in the yard. Our design can satisfy the old man's need. They like quiet, but they don't want to be isolated.
The roof in our design has a slope, it is good for the recycle of the rain and it is also the symbol of the traditional Chinese house. It is also the symbol of the traditional Chinese culture.

设计说明

以福建古代居住建筑——土楼为建筑原型，庭院作为生活的组织核心，本建筑利用建筑体量产生两个半封闭的庭院，并用廊道作为连接，彼此相通又满足老年人喜欢安静但又不与世隔绝的特点，屋顶采用坡屋顶，与中国古代的四水归堂相契合，同时也为建筑节能和太阳能设备提供了条件，在尊重文化地域性的基础上做到建筑节能。

泉州生态颐养服务中心设计

TOTAL FLOOR PLAN 1:650

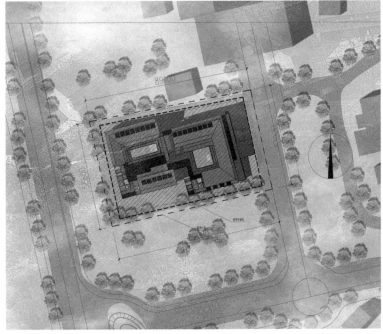

LIGHT SIMULATION

When building wai synthesis square courtyard, although increased building natural lighting, but the yard in ventilated effect is not good.

In order to increase the natural ventilation building, the building in two, increase the construction ventilation channel, courtyard ventilation effect is improved.

Considering the dominant wind direction, building staggered, increase architectural natural lighting and ventilation.

Windward side do back Taiwan processing for architecture, rich site landscape within the field of vision at the same time effectively improved the construction yard of natural lighting and ventilation.

SOUTH & EAST ELEVATION PLAN 1:200

光·栖院 LIGHT·HABITAT & YARD

泉州生态颐养服务中心设计

1st FLOOR PLAN 1:200

1 HEALTH ROOM 2 RECOVERY ROOM
3 INFIRMARY ROOM 4 VIEWING ROOM
5 THERAPEUTIC ROOM 6 EQUIPMENT ROOM
7 TREATMENT ROOM 8 PHYSICAL ROOM
9 SHOP 10 STAIRS & ELEVATORS
11 BARBER SHOP 12 BANK
13 READING ROOM 14 NETWORK ROOM
15 CHESS ROOM 16 CALLIGRAPHY STUDIO
17 MULTI-FUNCTION HALL 17-1 LOUNGE
18 GYM 19 DINING-ROOM
19-1 KITCHEN 20 WAREHOUSE
21 LAUNDRY ROOM 22 DUTY ROOM
23 RECEPTION ROOM 24 EQUIPMENT ROOM
25 STAFF ROOM 26 MEETING ROOM
27 ARCHIVE ROOM 28 OFFICE ROOM
29 OFFICE ROOM 30 DUTY ROOM
31 HALL 32 TOILET

- LIVING ROOM
- TRAFFIC
- MANAGEMENT SERVICES
- ROOMS
- MEDICAL AREA
- PUBLIC ACTIVITY AREA

LIGHT SIMULATION

Daylight factor
All the room lighting coefficient meets the requirements.

Natural light
All the rooms natural light is good.

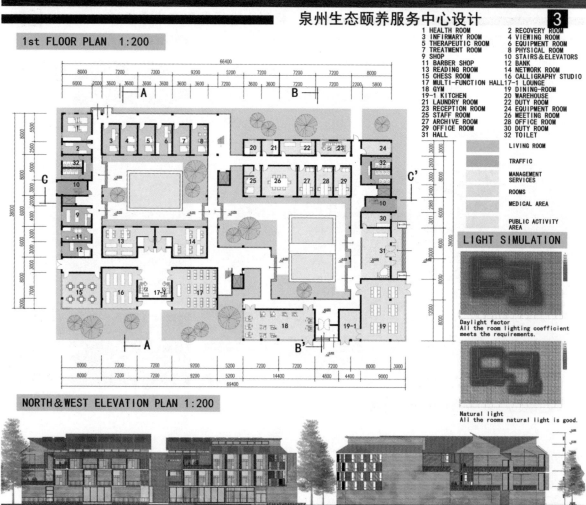

NORTH & WEST ELEVATION PLAN 1:200

光·栖院 LIGHT·HABITAT & YARD

WIND、SOLAR、HEAT ANALYSIS

Summer: The sun's elevation angle is large, class-rooms and cantilevered balconies cornices the south can effectively prevent the direct sunlight into the interior, reducing solar radiation and heat. The south high window grile reflect sunlight, thus blocking the light into interior.

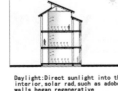

Daylight: Direct sunlight into the interior, solar rad, such as adobe walls began regenerative regenerator, through a direct biation into heat so that the classroom temperature; at the same time, benefit to the rooms in winter temperatures reach a comfortable room temperature requirements.

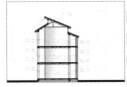

Winter: The winter direction northwest wind, the use of slopes and sheltered the role of the entire campus, reduce the external effects of wind on the rooms, use of solar hot chimney less role in indoor air circulation, the formation of ventilation.

OVERHEAD FLOOR

Empty space form of the building structure has the advantage of obvious, safe, moist insulation, ventilation, adjust measures to local conditions, a particular art style is stilt floor. is contribution to the human living.

架空层这种建筑结构形式的优势是明显的，安全、隔潮、通风、因地制宜、体现特定的艺术风情是架空层是对人类居住生活的贡献。

Winter: By the calculation of solar elevation angle, the balcony of the cornices doesn't blocking the light directly into the interior; at the same time, as both the south to the high window parallel with the sun angle, the light can be lauched into interior.

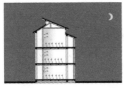

Night: In order to prevent a sharp fall in night temperature to make the rooms too much heat loss, heat to prevent students from accessing the next day when the temperature is too low, indoor temperature difficulties.

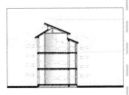

Summer: Summer wind direction for the weat wind, the use of slope and wind throught the campus, the role of the special position of the rooms window high pressure air to form, at the same time the use of draft to form a good indoor ventilation.

泉州生态颐养服务中心设计　　4

2nd FLOOR PLAN 1:200

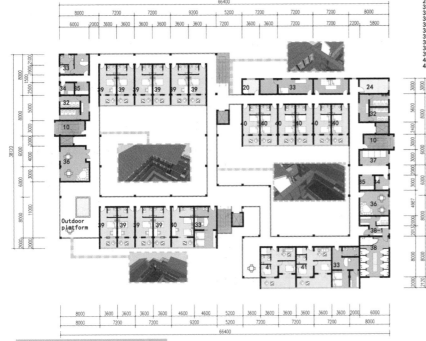

10 STAIRS & ELEVATOR
20 WAREHOUSE
24 EQUIPMENT ROOM
32 TOILET
33 NURSING STATION
34 GARBAGE ROOM
35 BOILED WATER ROOM
36 INTERACTION ROOM
37 LAUNDRY ROOM
38 MEN'S BATHROOM
38-1 DRESSING ROOM
39 SINGLE ROOM
40 DOUBLE ROOM
41 SUITE ROOM

- LIVING ROOM
- TRAFFIC
- MANAGEMENT SERVICES
- ROOMS

LIGHT SIMULATION

Daylight factor
All the room lighting coefficient meets the requirements.

Natural light
All the rooms natural light is good.

C-C' SECTION PLAN 1:200

 光·栖院 LIGHT · HABITAT & YARD

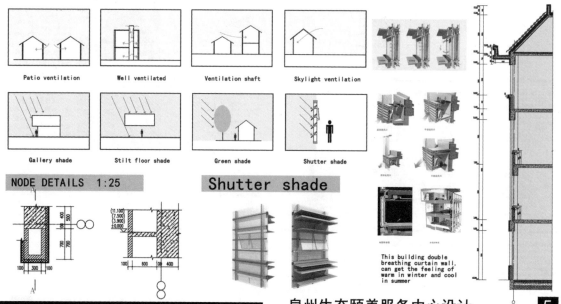

泉州生态颐养服务中心设计

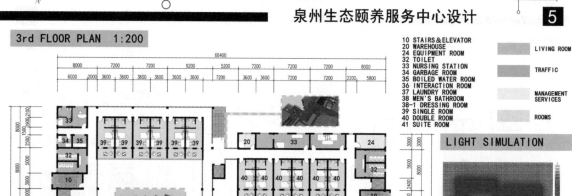

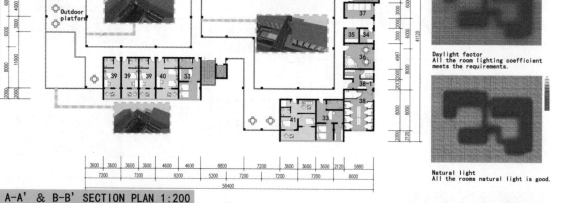

光·栖院　LIGHT · HABITAT & YARD

SOLAR THERMAL UTILIZATION

VERTICAL PLANTING

ECONOMIC INDICATORS

volume fraction	1.95
greening rate	55%
overall floorage	3136.6 ㎡
floor area	1608 ㎡
living room	15.78 ㎡/one
health care occupancy	3.4 ㎡/one
activity room	14 ㎡/one
set by the management service	4.41 ㎡/one

SOLAR PHOTOVOLTAIC

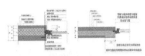

LOW-E GLASS

SPONGE CITY

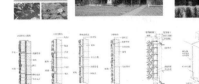
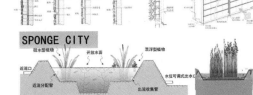

泉州生态颐养服务中心设计　6

economic analysis

	function		area	grass area
residence	single-bed room	18×28.8 ㎡	518.4 ㎡	1094.4 ㎡
	double room	14×28.8 ㎡	403.2 ㎡	
	double suite	3×57.6 ㎡	172.8 ㎡	
living / life auxiliary	public washroom	2	40 ㎡	80 ㎡
	public kitchen	1	26.4 ㎡	26.4
	public restaurant	1	112 ㎡	112 ㎡
	self-service laundry room	4	8 ㎡	32 ㎡	758.4 ㎡
	boiler room	4	8 ㎡	32 ㎡
	nursing station	8	42 ㎡	336 ㎡
	rubbish room	4	8 ㎡	32 ㎡
	exchanges hall	3	36 ㎡	108 ㎡
life service area	private bathroom	2	72 ㎡	132 ㎡
	barbershop	1	18 ㎡	
	barbershop	1	24 ㎡	
	post and telecommunications	1	18 ㎡	
health care occupancy / medical USES a room	dispensary	1	21.6 ㎡	129.6 ㎡
	observation room	1	21.6 ㎡	
	treatment room	1	21.6 ㎡	
	pharmacy	1	21.6 ㎡	
	office treatment	1	21.6 ㎡	
	psychotherapy room	1	21.6 ㎡	
health care occupancy	health room	1	28 ㎡	48 ㎡
	recovery Room	1	20 ㎡	
	psychological counseling room	0		
community room / activity room	reading room	1	57.6 ㎡	744.8 ㎡
	internet room	1	57.6 ㎡	
	chess and card room	1	56 ㎡	
	calligraphy studio	1		
	fitness center	1	84 ㎡	
	multi-purpose room	1	73.6 ㎡	
	the wind and rain gallery	△	360 ㎡	
set by the management servicem	General Duty Office	1	21.6 ㎡	21.6 ㎡	229.4 ㎡
	registration room	1	10 ㎡	10 ㎡
	office	2	21.6 ㎡	43.2 ㎡
	antechamber	1	21.6 ㎡	21.6 ㎡
	assembly room	1	43.2 ㎡	43.2 ㎡
	muniment room	1	21.6 ㎡	21.6 ㎡
	washhouse	1	10 ㎡	10 ㎡
	the worker housing	1	21.6 ㎡	21.6 ㎡
	repair parts stock	1	15 ㎡	15 ㎡
	facilities room	1	21.6 ㎡	21.6 ㎡

地源热泵是利用水与地能（地下水、土壤或地表水）进行冷热交换来作为地源热泵的冷热源，冬季把地能中的热量"取"出来，供给室内采暖，此时地能为"热源"；夏季把室内热量取出来，释放到地下水、土壤或地表水中，此时地能为"冷源"。

综合奖·三等奖
General Prize Awarded · Third Prize

注 册 号：4636
项目名称：半院（西安）
　　　　　Half of Yard（Xi'an）
作　　者：车喜刚、雷宸骁、刘雨龙、
　　　　　陈函璐、何静怡、成　侃
参赛单位：西安科技大学
指导老师：孙倩倩

专家点评：

作品单元综合了养护、餐厅和住宿功能，具有流线便捷、服务半径短的较突出优点。通过有效的剖面设计，较合理地解决了基地南北山形的高差问题。立面材料采用夯土本地材料等工法，较大面积地采用太阳能板，符合本次竞赛主题的要求。太阳能与建筑一体化融合不够，南北流线稍显长。

The project of the work, integrating the function of maintenance, restaurants and accommodation, embodies relatively outstanding features of convenient streamline and short service radius. The effective design of its profile can solve the problem of altitude difference of south-north mountains in the base in a relatively reasonable manner. The work adopts rammed earth as elevation materials, and solar panel in a relatively large areas, conforming to the requirements of this competition theme. However, the integration of solar energy and building is not conducted well, and the south-north streamline seems rather longer.

半 院 HALF OF YARD 02

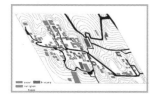   

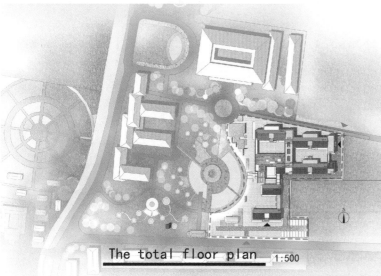

The total floor plan 1:500

 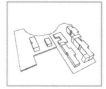

 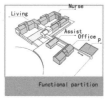

Axis and yard | Functional partition | The summer wind | The winter wind

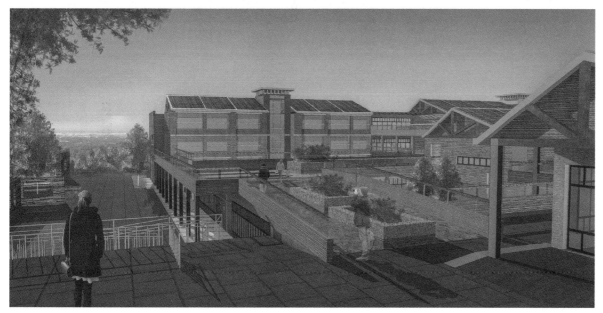

半 院   HALF OF YARD   03

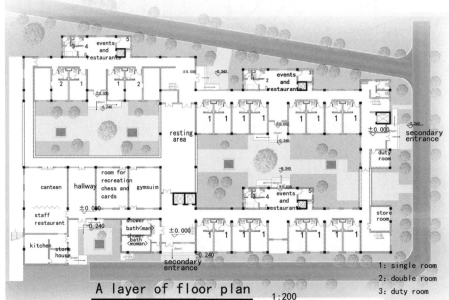

A layer of floor plan　1:200

1: single room
2: double room
3: duty room
4: nurse station
5. store room

Barrier-free design

Lighting equipment
Security considered to ensure enough brightness

The decoration of the ground
Application materials, prevent slippery, prevent tripping.

Lighting equipment
Security should be considered to ensure enough brightness.

The armrest (measures to prevent falling)
Should be on the height of more than 1100 mm above ground.

Armrest (movements and walking auxiliary)
Set at least one side Set on the height of the above ground, 700~900 mm.

Vertical height difference
The ground structure without vertical height difference.

Inward and outward width
Effective width 1000 mm

The vertical height difference of inward and outward.
The ground structure without vertical height difference adopted.

It is set the footlights,
Avoid shadow step plate appear.
But should avoid light directly into his eyes.

Accept
In ensuring that appropriate capacity, but also set in effortless can take place.

The decoration of the ground:
Application materials, prevent slippery, prevent tripping.

The vertical height difference of inward and outward:
Structure without vertical height difference adopted

Emergency device handrail
For the convenience of squatting, set up handrail

The decoration of the ground
Application materials, prevent slippery, prevent tripping.

The west facade　1:200

半院　HALF OF YARD　04

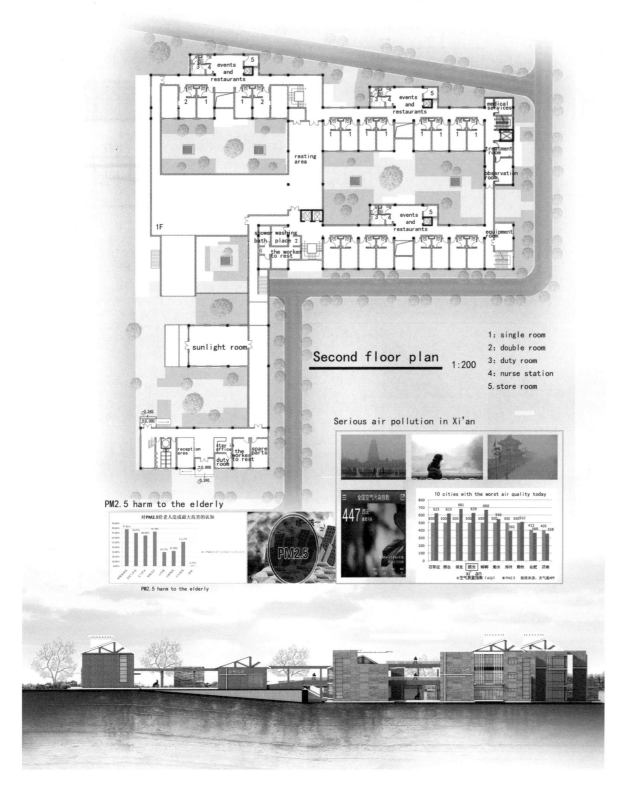

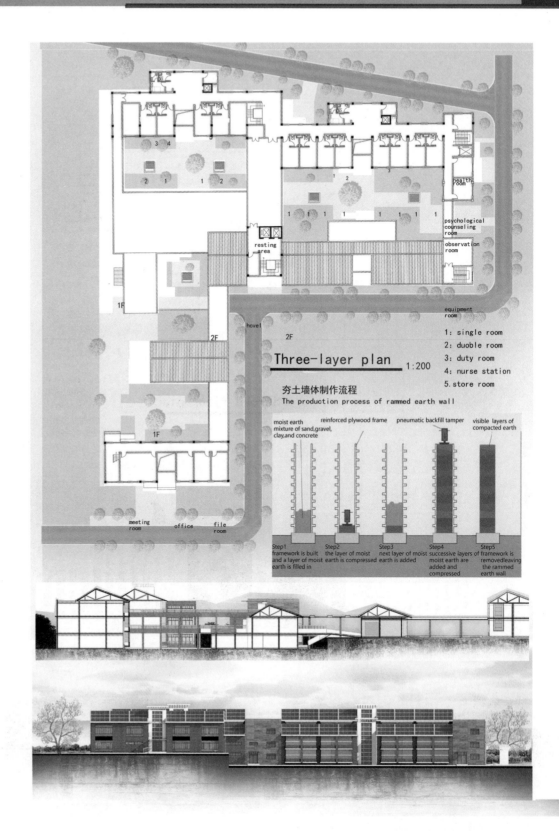

半院 / HALF OF YARD

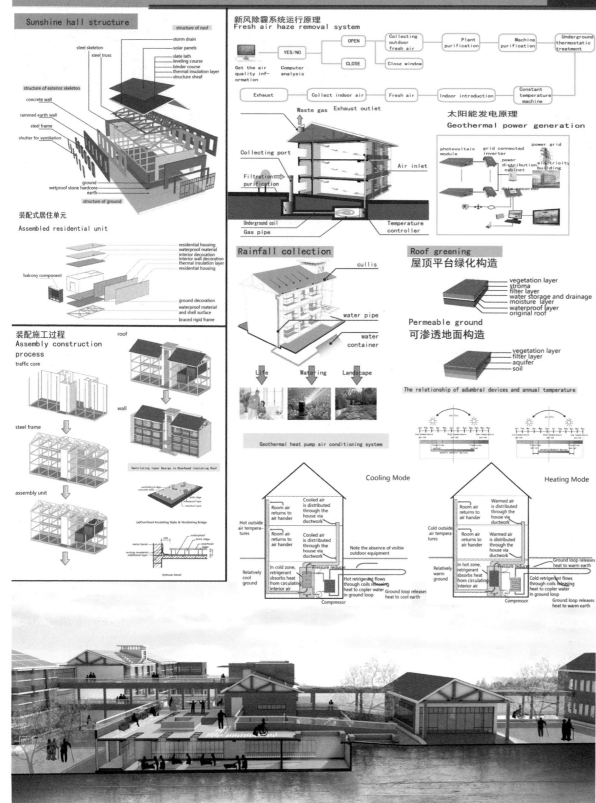

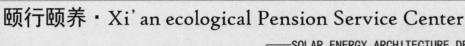

综合奖·三等奖
General Prize Awarded · Third Prize

注 册 号：4909
项目名称：颐行颐养（西安）
　　　　　Roaming in the Sunlight (Xi'an)
作　　者：吕丹妮、江海华、武学志
参赛单位：华中科技大学
指导老师：徐　燊

专家点评：
作品总平面布局基本适应地形条件，并充分预留了二期发展用地。利用与斜坡屋顶的结合，太阳能与建筑一体化设计具有一定的特点。条状的建筑布局略显单调，建筑主入口交通流线尺度和空间有待处理。

The overall plane layout of the project adapts to topographic conditions basically, and makes full reservation for the second-stage land use for development. With the utilization of sloping roof, the design integrating solar energy and building embodies a certain feature. However, the bar-shaped architectural composition seems more monotonous, and it remains to be addressed for scale and space of traffic streamline of the building's main entrance.

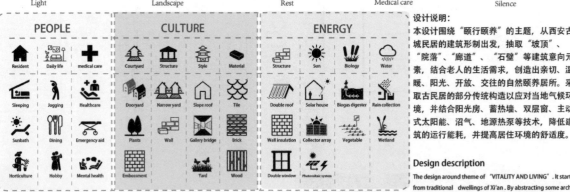

设计说明：
本设计围绕"颐行颐养"的主题，从西安古城民居的建筑形制出发，抽取"坡顶"、"院落"、"廊道"、"石壁"等建筑意向元素，结合老人的生活需求，创造出亲切、温暖、阳光、开放、交往的自然颐养居所。采取古民居的部分传统构造以应对当地气候环境，并结合阳光房、蓄热墙、双层窗、主动式太阳能、沼气、地源热泵等技术，降低建筑的运行能耗，并提高居住环境的舒适度。

Design description
The design around theme of "VITALITY AND LIVING". It starts from traditional dwellings of Xi'an. By abstracting some architectural intent elements, such as "Sloping roof", "Coutyard", "Gallary bridge", "Embossment", etc., and combining the needs of the livelihood of the elderly, we intend to creat a familiar environment for older people which may make it easy for them to adapt their new home and a natural home which is warm, sunny, open and comfortable. To deal with local climate and environment, we adopt some construction of traditional dwellings. Besides, we combined those strategies with the "solar house", "solar wall", "photovoltaic system", etc., to reduce the consumption of energy and provides more comfortable environment.

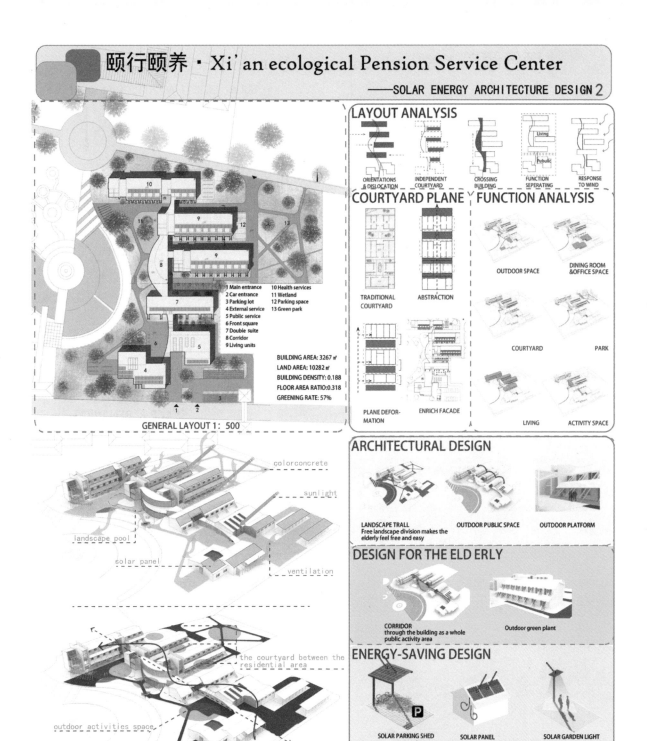

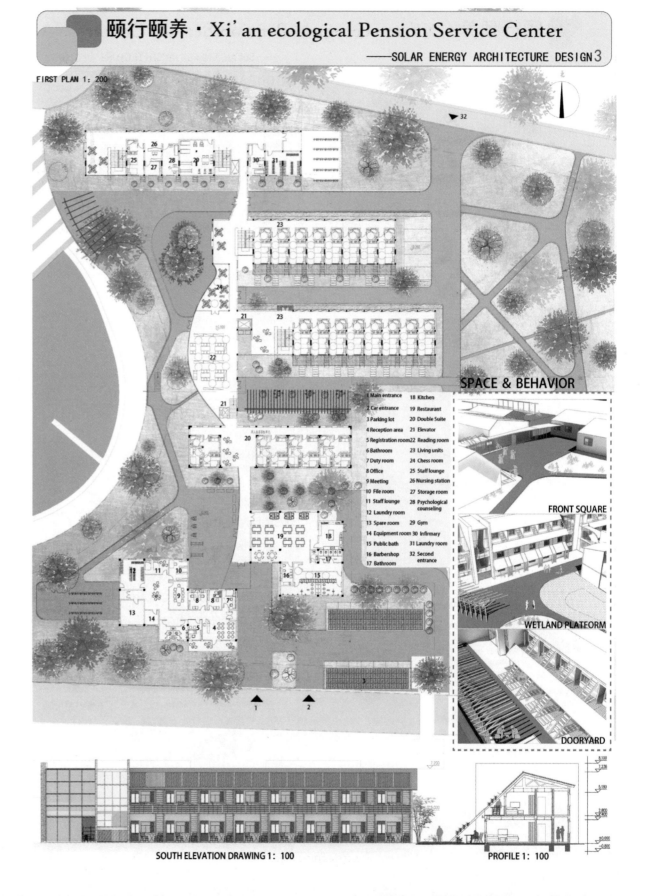

颐行颐养 · Xi'an ecological Pension Service Center
——SOLAR ENERGY ARCHITECTURE DESIGN 4

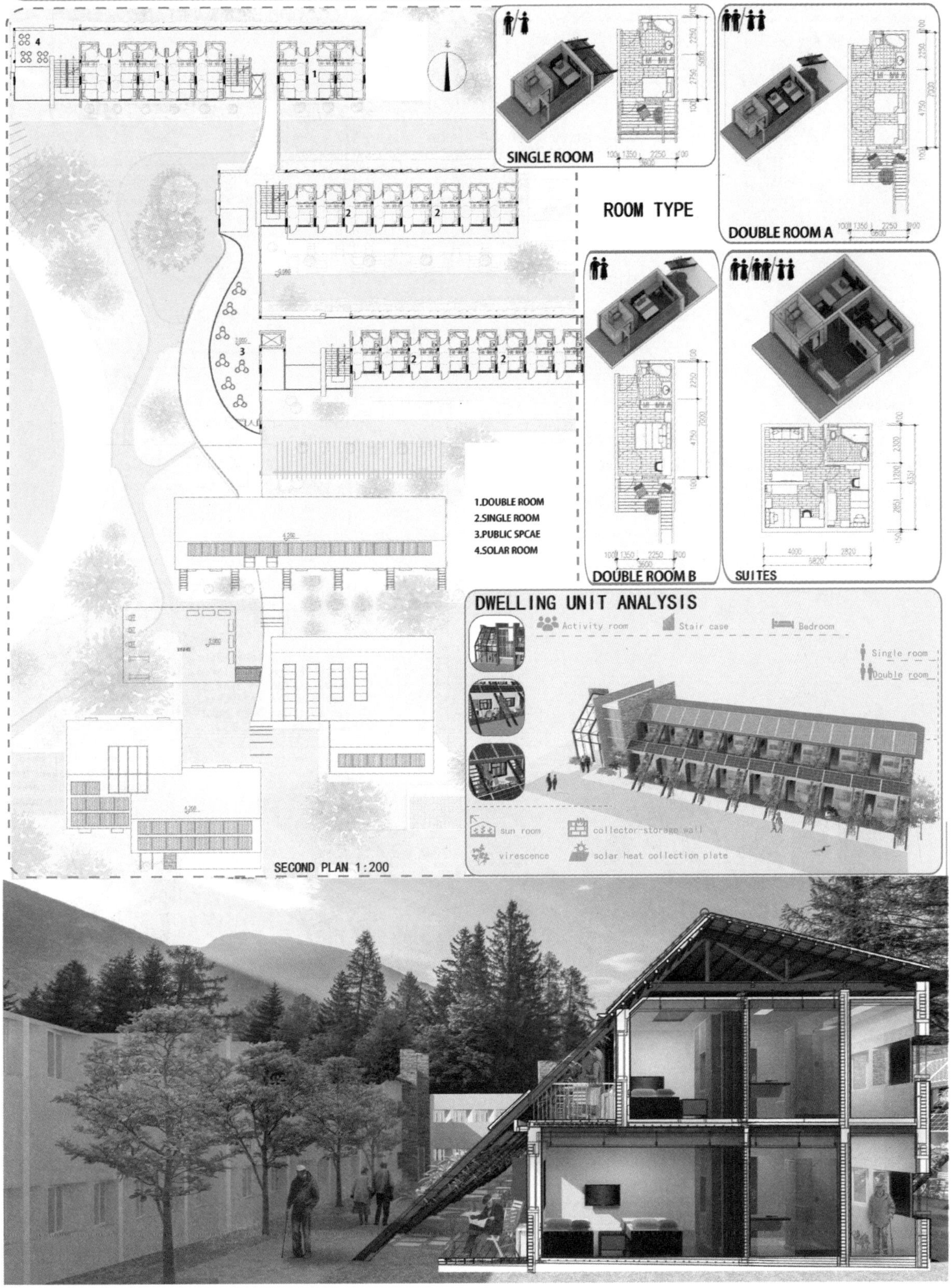

ROOM TYPE
- SINGLE ROOM
- DOUBLE ROOM A
- DOUBLE ROOM B
- SUITES

1. DOUBLE ROOM
2. SINGLE ROOM
3. PUBLIC SPCAE
4. SOLAR ROOM

SECOND PLAN 1:200

DWELLING UNIT ANALYSIS
- Activity room
- Stair case
- Bedroom
- Single room
- Double room
- sun room
- collector-storage wall
- virescence
- solar heat collection plate

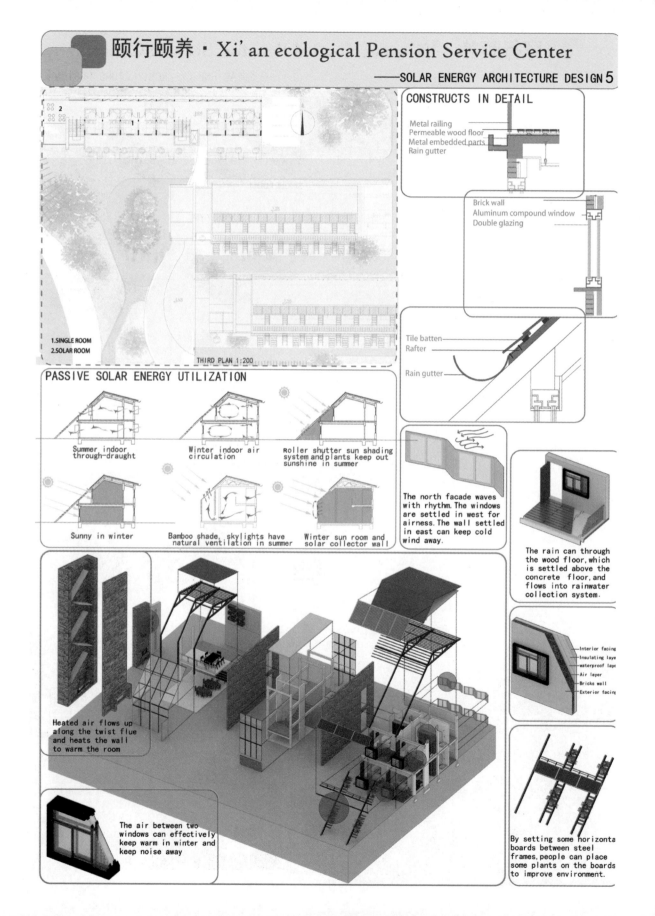

颐行颐养 · Xi'an ecological Pension Service Center
—— SOLAR ENERGY ARCHITECTURE DESIGN 6

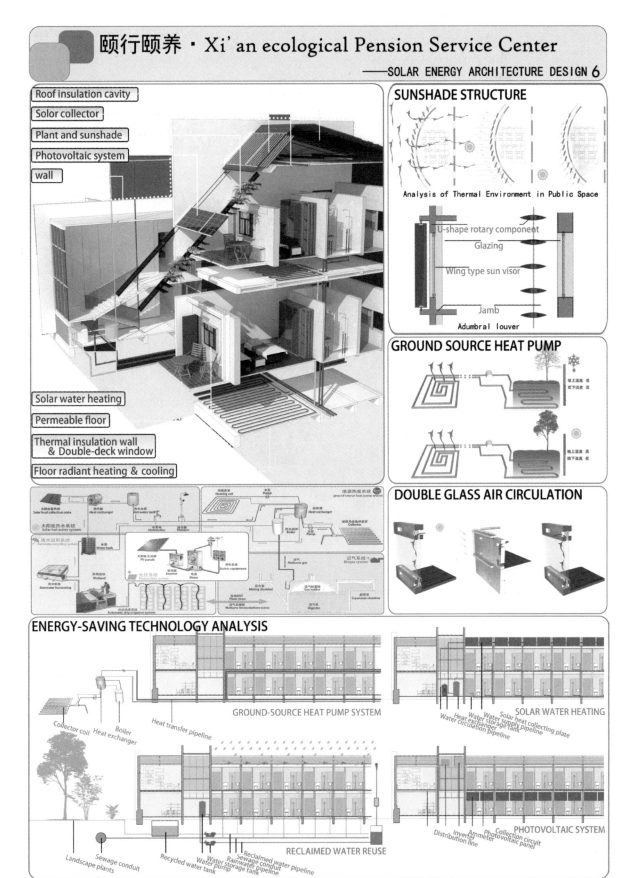

综合奖·三等奖
General Prize Awarded · Third Prize

注 册 号：4988
项目名称：土生土长（西安）
　　　　　Half-underground Community
　　　　　（Xi'an）
作　　者：张军军、张宇涛、王冠军
参赛单位：东南大学
指导老师：张　宏

专家点评：

作品与其他方案完全不同的独特构思，植根于西北传统的生土建筑，符合当地气候特点，每一单元由四个独立房间组成，既具有独立性和院落，也可以共享中间阳光暖房，非常适合老年人的生活需求。作品未考虑二期建设发展用地，单元的公共连廊缺乏识别性和公共交流空间。

Thoroughly different from other projects with its unique designing concept, the work sets its root in earth buildings of northwest tradition, which accords with local climate features. Each unit, composed of four separate rooms, embodies the independence and its yards, and can also share warming residence in the middle, which is suitable for the needs of elderly life. However, the project fails to take into consideration the second-stage land use for development, and the public corridors of units lack identifiability and space for social interaction.

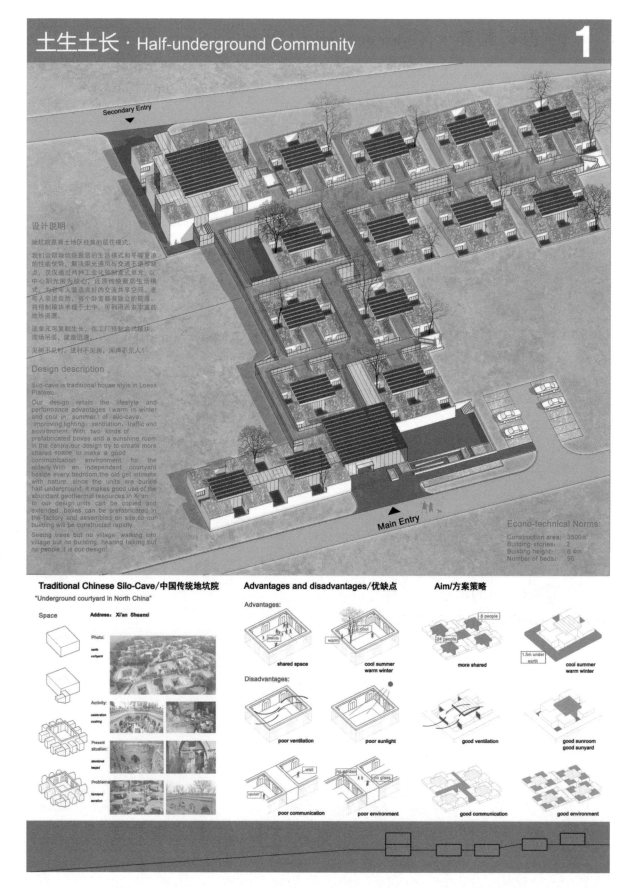

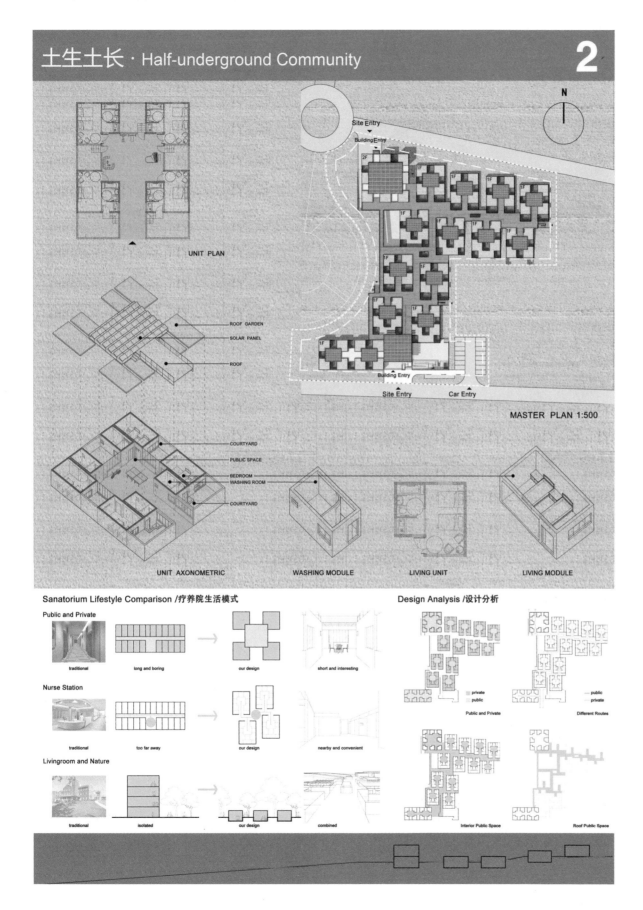

土生土长 · Half-underground Community

3

Living Core/中心起居室

Roof Garden/屋顶花园

First Floor 1:250

Section A-A 1:250

土生土长 · Half-underground Community

4

2017 台达杯国际太阳能建筑设计竞赛获奖作品集

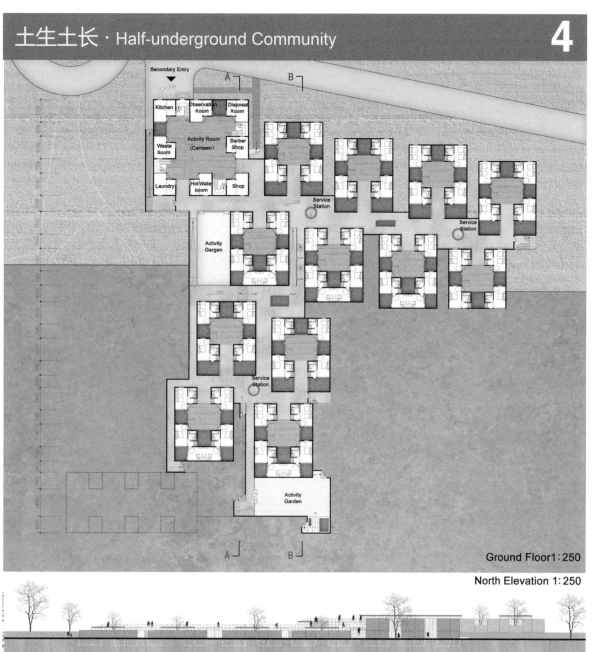

Ground Floor 1:250

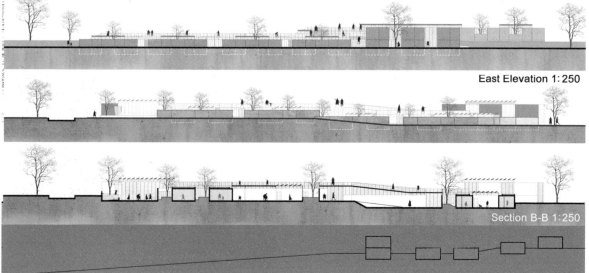

North Elevation 1:250

East Elevation 1:250

Section B-B 1:250

051

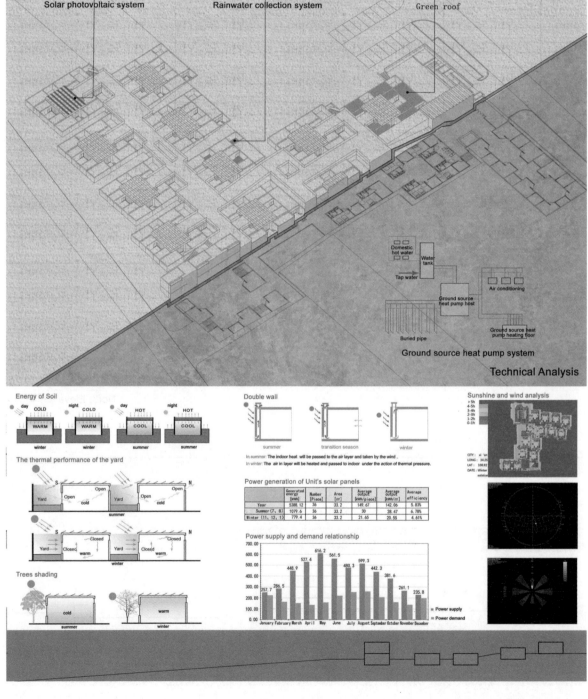

土生土长 · Half-underground Community

6

2017 台达杯国际太阳能建筑设计竞赛获奖作品集

進村不見房　聞聲不見人　見樹不見村

Industrialized Construction/工业化建造

1. Prefabricated concrete module
2. Light steel structure
3. Environmental improvement

Integrated Product /集成产品

- Prefabricated box building module
- Light steel frame structure
- Solar integrated technology
- Roof greening integration technology
- Integral bathroom and kitchen
- Ground source heat pump
- Water circulation heating
- Rainwater recollection technology

Extendibility/可扩展性

- Basic unit
- Our building
- Extended building

053

综合奖・三等奖
General Prize Awarded · Third Prize

注 册 号：5165
项目名称：暮·光（西安）
　　　　　Elderly in Sunshine (Xi'an)
作　　者：杨　娇、陈其龙、刘　颖、
　　　　　衣志浩
参赛单位：西安建筑科技大学、
　　　　　沈阳建筑大学
指导老师：张　群、赵西平

专家点评：

作品以四个合院串联而成，所有老人单元南向有阳台，可防北风而保暖，平面联络清晰有致，动线明确，平面配置良好。本设计对于老人空间的行为分析十分周到，南方取阳光采暖。另外，本设计采取西安青瓦土墙造型，对风土文化有所反映，值得赞许。对于未来扩建的可能性作品考虑不足。本设计虽有地下雨水利用设计，但对于基地北低南高的自然雨水储存利用考虑不足。

The work is cascaded by four courtyards, and all the units oriented towards the south for the elderly have balconies which can keep them warm by blocking north wind. The work boasts clear plane connection, specific generatrix and fine plane configuration. This design shows full consideration to behavior of the elderly, and it is easier to collecting heat from the sun in the south. In addition, it is worthy of praising that the design's adopting the Xi'an moulding of grey tile and cob wall is responsive to local customs and culture. However, the work gives less consideration to possibility of further extension. Besides, it fails to fully consider the storage and use

Site Location Analysis

Information
Location: Taigou Village, Ziwu Town, XI'an
　　　　　N 34°02′, E 108°53′
Climate: Sub-humid warm temperate continental monsoon climate
Lowest/Highest elevation: 384.7m/2886.7m

Design Specification:
The design theme of nursing home is making full use of the sun. The building space design bases in traditional Shaanxi Province Courtyard, in order to make the elderly return to traditional life and fully embrace the sunlight. On the active design of solar energy, Lay solar collector on building south slope roof and join solar panel in south glass of building. On passive solar design, make the sun into the greenhouse space, heat the indoor air, and then make the rock wool in the floor heat storage. Set high side window in the building, so that make the floor strata, and the wall heat storage. Using sustainable green energy saving design, driving the development of the aged service industry.

该颐养服务中心设计以充分利用阳光为出发点，建筑设计以陕西地区传统院落为原型，形成秩序中富有变化的建筑体量。太阳能主动式设计方面，在建筑南向坡屋面铺设太阳能集热器，在建筑南向的玻璃夹层中加入窗户集热板。太阳能被动式设计上，使太阳光进入温室空间，加热室内空气，进而使地板蓄热；建筑中的高侧窗使得阳光得以进入走道空间，从而使楼地层和墙体蓄热。运用可持续的绿色节能设计，带动当地的养老产业发展。

1 暮·光
Elderly in Sunshine
Status Analysis

Climate Analysis

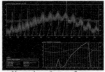
Hourly data for hottest day (average)

Weekly data

Best orientation

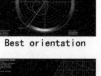
Wind analysis　Solar radiation　Psychrometry

Traditional Residence

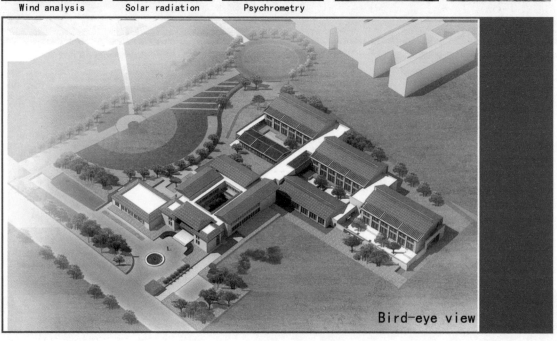
Bird-eye view

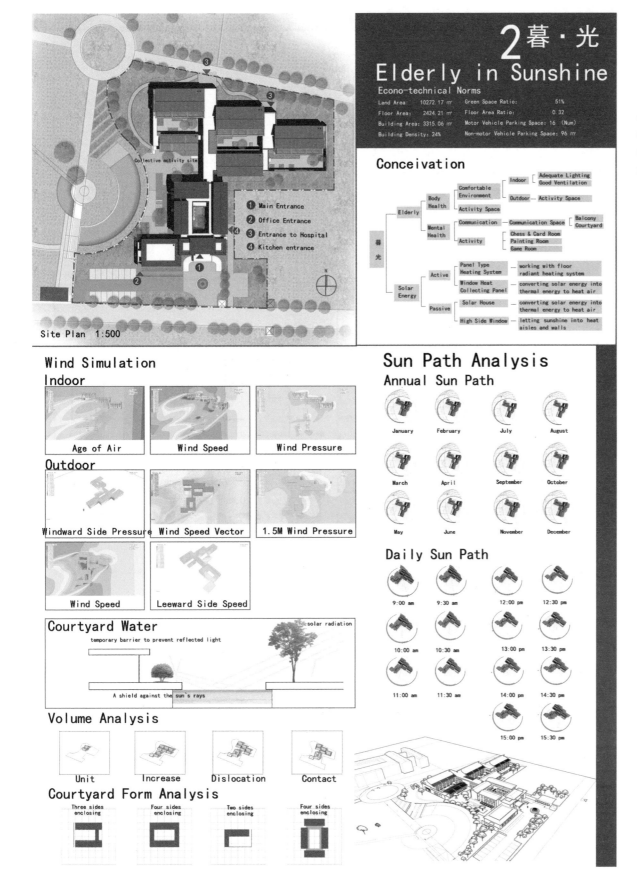

of natural rainwater from the bases featuring altitude in the south higher than that in the north despite the design of underground rainwater utilization.

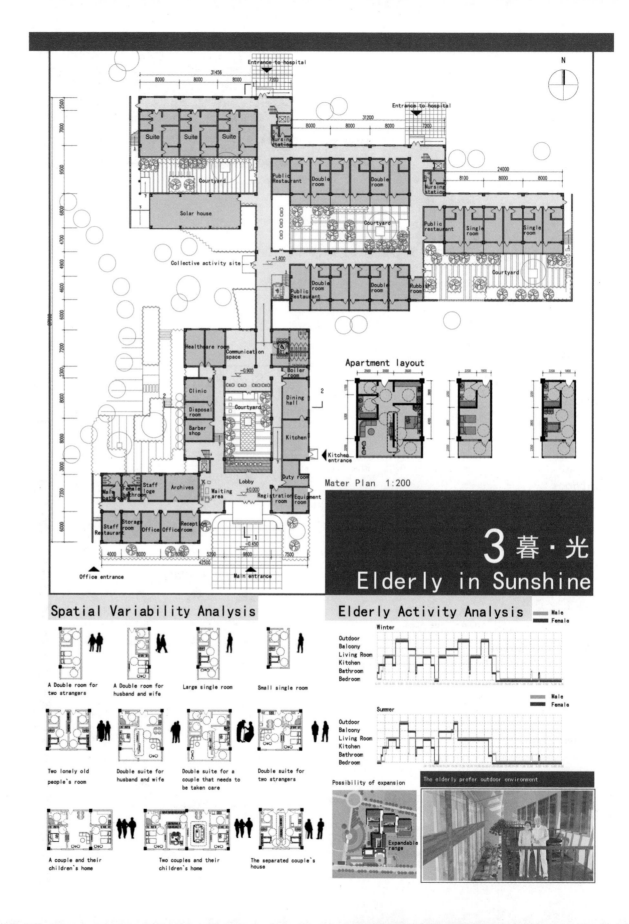

4 暮·光
Elderly in Sunshine

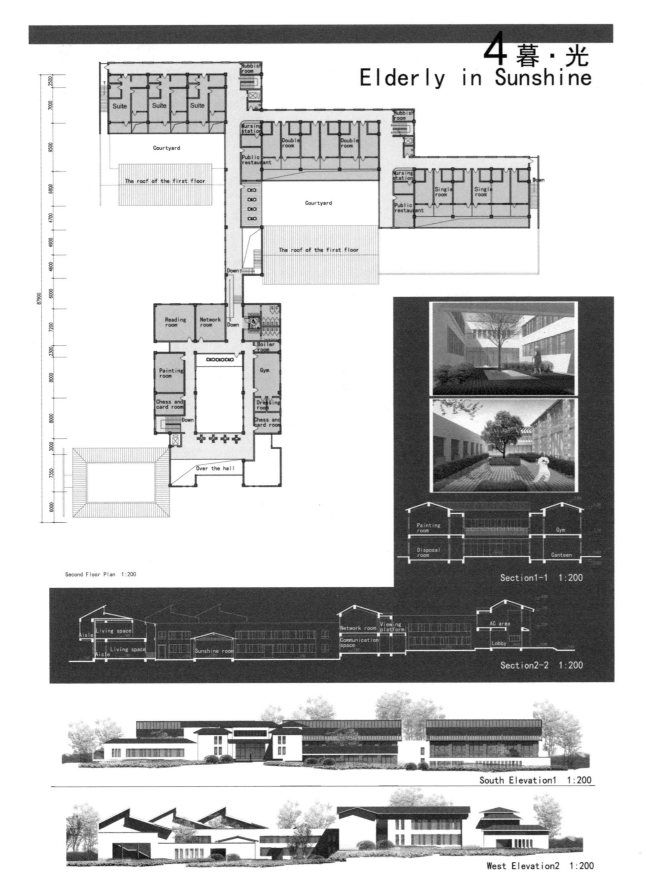

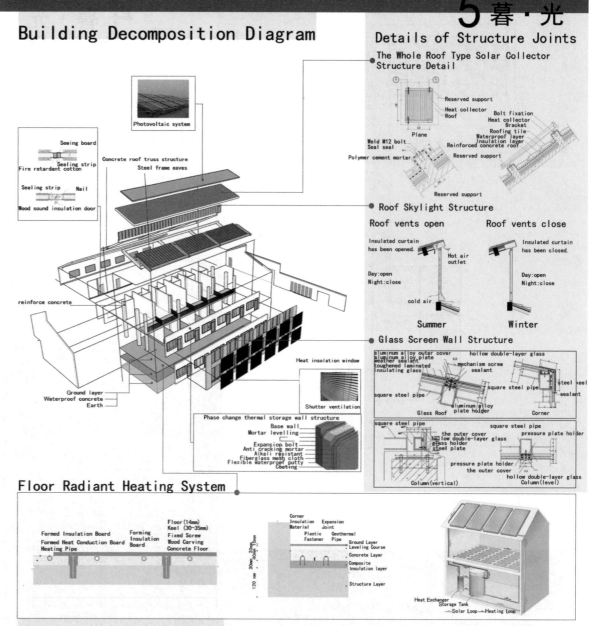
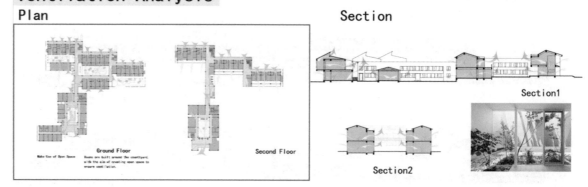

6 暮·光
Elderly in Sunshine

Analysis Of Solar Energy Technology In Living Space

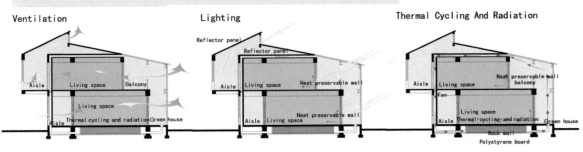

Daylighting Analysis In Different Space

Section Perspective

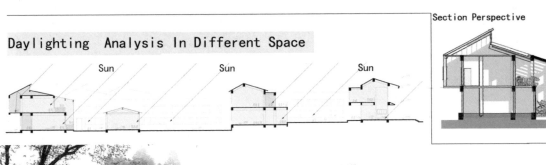

Rainwater Utilization

诗意·栖居 1 POETIC DWELLING

综合奖·三等奖
General Prize Awarded ·
Third Prize

注 册 号：5259
项目名称：诗意栖居（西安）
　　　　　Poetic Dwelling（Xi'an）
作　　者：曹赓、郑欣欣、林佳昕
参赛单位：广州大学
指导老师：李丽、刘源

专家点评：

作品为四平八稳、机能完善的设计，但把基地占领殆尽，对未来扩建发展不利。本设计南北距离太远，对老人行为不利，同时，两处入口前面停车与活动空间预留不足，使用机能不足。本设计与太阳能光伏或被动式太阳能之一体设计不足，绿色设计之造型不能凸显。本设计合院配置对防患东北季风入侵并无良好反映，对于老年住宅的保暖设计不足。

The design is characterized by stable structure and impeccable function. However, it is unfavourable for further extension and development to occupy all the bases. It develops long distance between the south and north, which is inconvenient for behavior of the elderly. Meanwhile, less space is reserved for parking and activities in front of the two entrances, lacking capability of more use. The design fails to demonstrate its sufficient integration with solar photovoltaic or passive solar energy, and fails to highlight moulding of green design. Moreover, the courtyard configuration can't prevent northeast monsoon from invading buildings, which lacks effectiveness in the design to maintain residence of the elderly warm.

■ **Daylighting analysis**

 Using double roof, using the double roof layer between the cooling air, avoid the top roof directly heated indoor temperature is too high

Heat insulation curtains

 Summer day: during the day will be reflective insulation curtain is placed on one side of a close to the glass, open the glass hole at the top and bottom, Window with the curtain in the middle of cold and hot air cycle, the sunlight room air temperature have no obvious rise.

 Winter day: the sun radiation into the sunlight, temperature, air vents on the adjacent wall, transfer heat to the room, the room air through a wall of vents Yang.

 Summer night: removing insulation curtains and wall body absorb heat radiation from indoor to outdoor, open the window, at the same time, in the form of natural ventilation cooling outward.

 Winter night: close the vents and use the up and down insulation curtain, prevent indoor to outdoor heat loss.

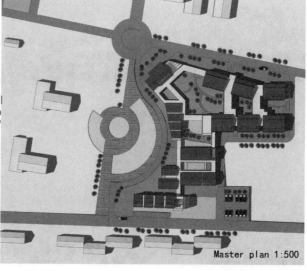

Master plan 1:500

■ **Wind environment simulation analysis**

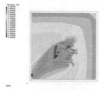

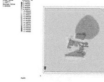

Summer: according to the Chinese building thermal environment analysis special meteorological data set "construction simulation summer conditions surrounding flow field distribution, set the direction for the NE (north east 45°), wind speed 2.3 m/s.

Winter: according to the Chinese building thermal environment analysis special meteorological data set "construction simulation summer conditions surrounding flow field distribution, set the direction for the NE (north east 45°), wind speed 1.6 m/s.

■ **Radiation analysis**

Wind: building around the pedestrian flow field distribution, basic no vortex formation, the overall ventilation in good condition. Pedestrian area around 1.5 m height, the wind speed is less than 0.6 m/s basic, overall, less than 5 m/s, wind velocity amplification coefficient is less than 2, conform to the requirements of the pedestrian comfort.

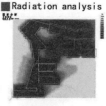

Wind: building around the pedestrian flow field distribution, basic no vortex formation, the overall ventilation in good condition. Pedestrian area around 1.5 m height, the basic wind speed is less than 1 m/s, as a whole is less than 5 m/s, the wind velocity amplification coefficient is less than 2, conform to the requirements of the pedestrian comfort.

诗意·栖居 2 POETIC DWELLING

■ Envelope Insulation

■ Thermal analysis

■ Geothermal energy utilization

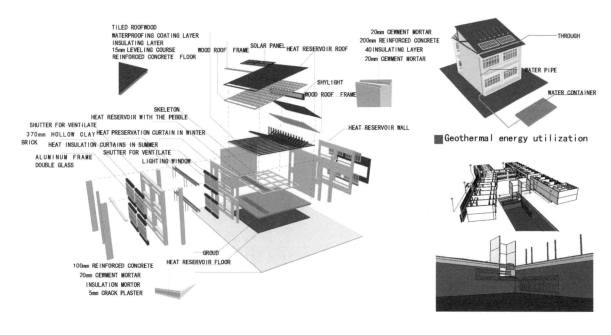

■ The room detail analysis

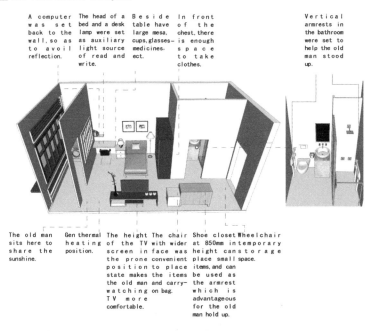

A computer was set back to the wall, so as to avoid reflection.

The head of a bed and a desk lamp were set as auxiliary light source of read and write.

Beside table have large mesa, cups, glasses-medicines ect.

In front of the chest, there is enough space to take clothes.

The old man sits here to share the sunshine.

Gen thermal heating position.

The height of the TV with wider screen in face was the prone convenient position to place state makes the items the old man and carry-watching on bag. TV more comfortable.

The chair with wider screen in face was convenient to place items, and can be used as the armrest which is advantageous for the old man hold up.

Shoe closet Wheelchair at 850mm intemporary height canstorage place small space.

Vertical armrests in the bathroom were set to help the old man stood up.

Profile control perspective

诗意·栖居 3 POETIC DWELLING

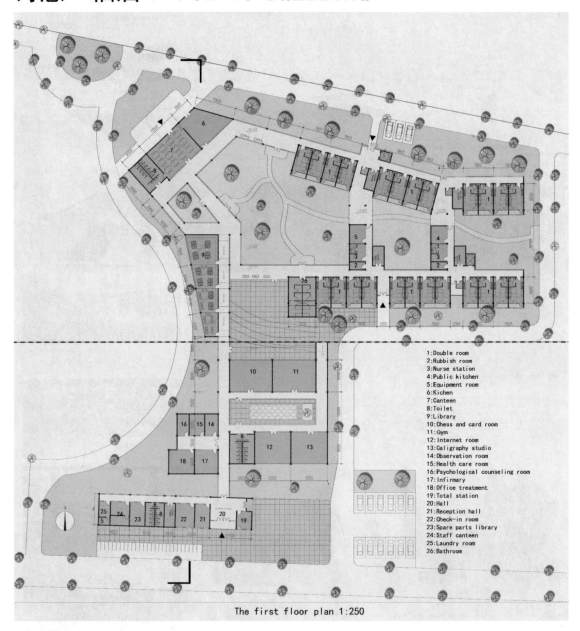

The first floor plan 1:250

1: Double room
2: Rubbish room
3: Nurse station
4: Public kitchen
5: Equipment room
6: Kitchen
7: Canteen
8: Toilet
9: Library
10: Chess and card room
11: Gym
12: Internet room
13: Caligraphy studio
14: Observation room
15: Health care room
16: Psychological counseling room
17: Infirmary
18: Office treatment
19: Total station
20: Hall
21: Reception hall
22: Check-in room
23: Spare parts library
24: Staff canteen
25: Laundry room
26: Bathroom

South Elevation 1:250

North Elevation 1:250

诗意·栖居 4 POETIC DWELLING

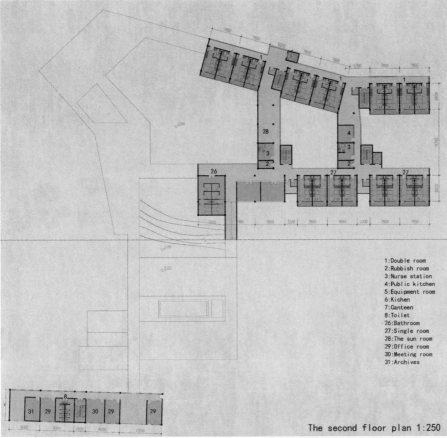

1: Double room
2: Rubbish room
3: Nurse station
4: Public kitchen
5: Equipment room
6: Kitchen
7: Canteen
8: Toilet
26: Bathroom
27: Single room
28: The sun room
29: Office room
30: Meeting room
31: Archives

The second floor plan 1:250

设计说明：
建筑位于西安市子午镇台沟村，夏季炎热多雨，冬季寒冷，在本设计中，利用太阳能，做到夏季隔热，冬季保温。同时，注意老人使用尺度，符合康健老人和失能老人共同使用的要求。本设计主要关注老人的休憩空间，每一个房间单元设有阳光间，结合西安地区特有的气候，达到冬天保温、夏季隔热的要求。同时，建筑具有丰富的活动空间，利于老人进行活动。

Design description:
Building in Xi'an city taigou village with cold winter, summer heat and rainy. in this design, the use of solar energy, do summer heat insulation and winter heat preservation. At the same time, pay attention to the old man use scale, conform to fit the old man and the common use of disability, old man. This design mainly focuses on the old man's leisure space, each unit has the sun room, combining with the characteristic of the climate in Xi'an region, to achieve heat preservation in winter, summer heat insulation requirements. Construction has a wealth of space at the same time, good for the old man.

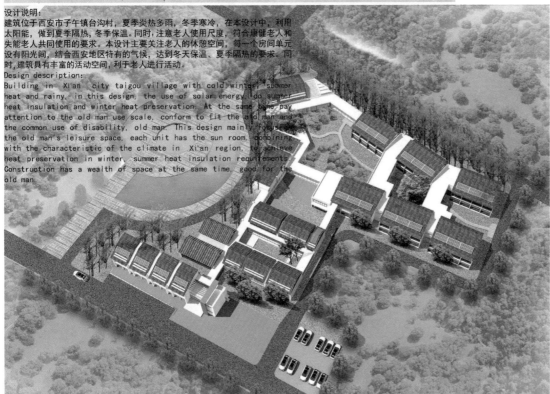

综合奖·三等奖
General Prize Awarded · Third Prize

注　册　号：5696
项目名称：安养享阳（泉州）
　　　　　Enjoy Sunshine Convalesce
　　　　　(Quanzhou)
作　　　者：陈艺松、刘来凤、赵航宇、
　　　　　岳开云
参赛单位：石家庄铁道大学
指导老师：高力强、何国青

专家点评：

作品建筑平面为两个相互联系的庭院，建筑功能完整，流线合理，适宜老年人生活并可满足其医疗需求。太阳能主、被动技术应用合理，设备转换传输效率高，建筑可实施性强，技术经济指标合理。但建筑三层两部电梯竖向交通较为困难，围合后自然通风略显不足，与南方建筑气候协调不够。

The building has two interrelated courtyards with integral function and reasonable streamline, which is conducive to demands of the elderly for life and medical care. What's more, the work enjoys rational application of active and passive technologies, high efficiency in equipment transformation and transmission, strong feasibility in construction, and reasonable index for technology and economy. Nevertheless, it is rather difficult for the two elevators at the third floor of the building to implement vertical transportation strategies, the natural ventilation after the enclosure rather fails to exert more effect, and it is insufficient to coordinate with climate in the south.

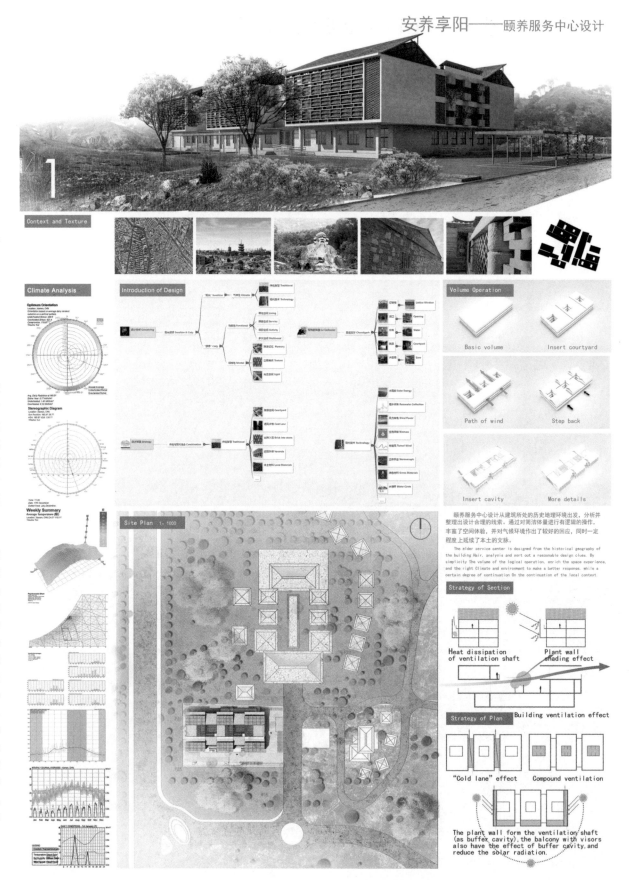

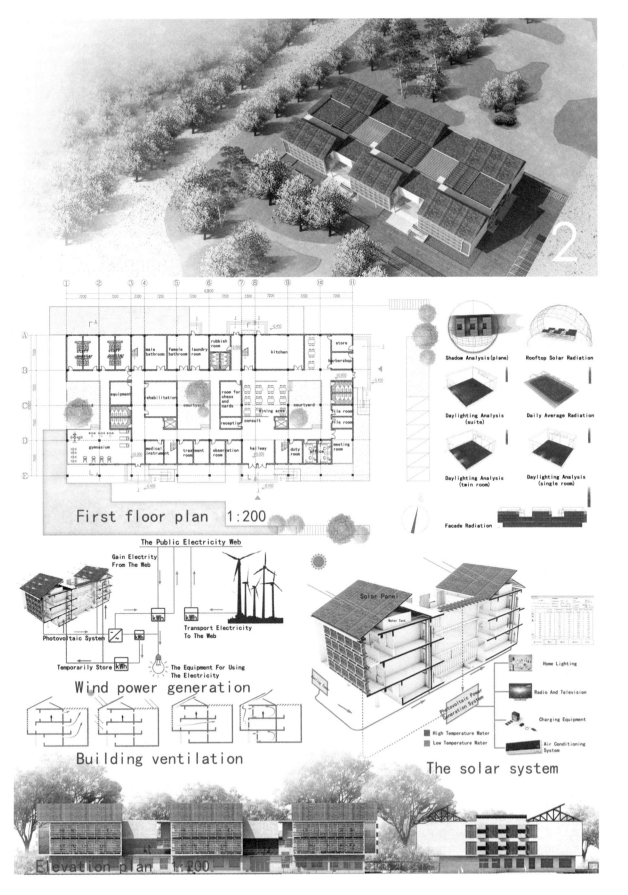

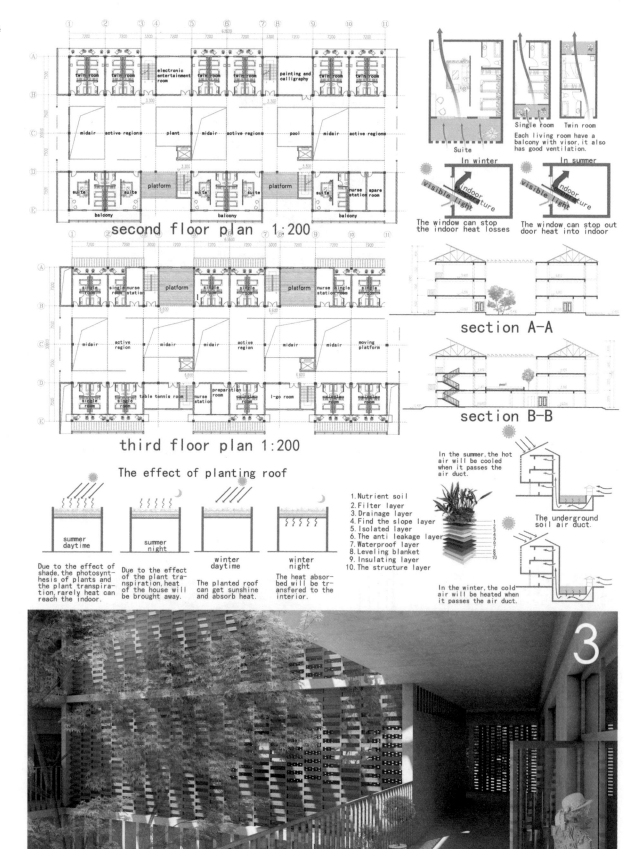

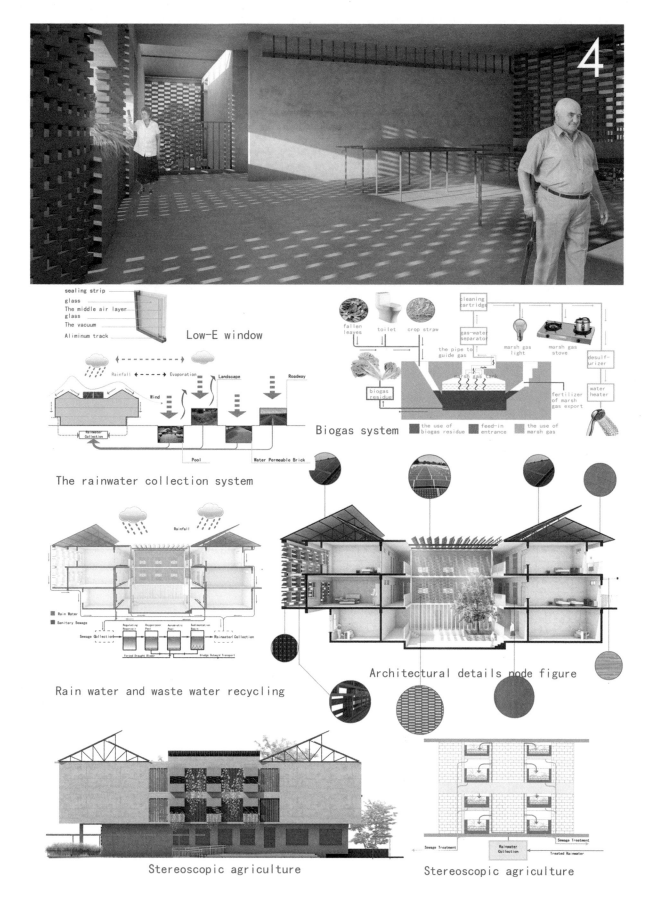

综合奖·优秀奖
General Prize Awarded · Honorable Mention Prize

注 册 号：4637
项目名称：颐园——老城墙下的怡养空间
　　　　　（西安）
　　　　　Home for the Old（Xi'an）
作　　者：王兰萍、张露露、张　雁、
　　　　　杨　吉、安建辉
参赛单位：西南科技大学
指导老师：白　雪、赵　祥、董美玲

1 颐园——老城墙下的怡养空间
HOME FOR THE OLD

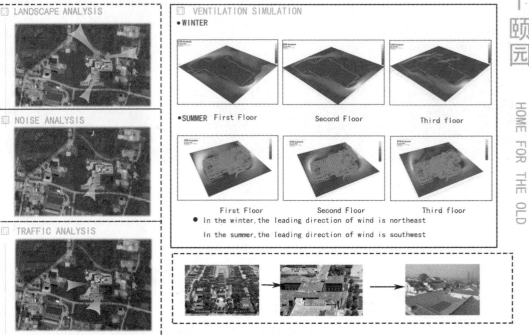

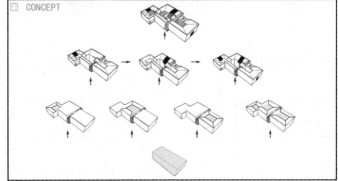

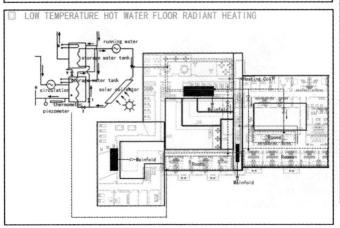

DESIGN SPECIFICATION

设计说明：正所谓，老有所依。老人对于生活和家庭的归属感都是极其强烈的，随着世界的发展，老人的生活更加应该受到重视。轻度失能的老人，基本健康的老人，活泼的老人，针对不同的颐养类型，让他们感到温暖和关怀，这也是我们一直追求的，意在为老人群体们打造一个健康颐养的生活环境。建筑利用了太阳能烟囱的被动太阳能利用技术来解决夏季通风，利用集热蓄热墙体和阳光间等被动太阳能技术来解决冬季供暖。供暖方式为集中供暖，利用了低温热水辐射地板的主动太阳能利用技术。同时，分析西安的气候条件，考虑太阳高度角和太阳方位角，在"四水归堂"的创意上更加深入地解决方案的夏季通风和冬季供暖问题。

We always hear that olds should have their attribution. The old people's sense of belonging is so strong. With the development of the world, the old people's life should be valued. Mild disability in the elderly, basic health old man, and active man, as for different types of care, our dream is to make they feel warm and cared. We give the design scheme name that is "YI YUAN".

Our purpose is to create a health life environment, we use passive solar and active solar to design reasonable elevation with considering solar altitude. We would solve problem of ventilation in summer and heating in winter.

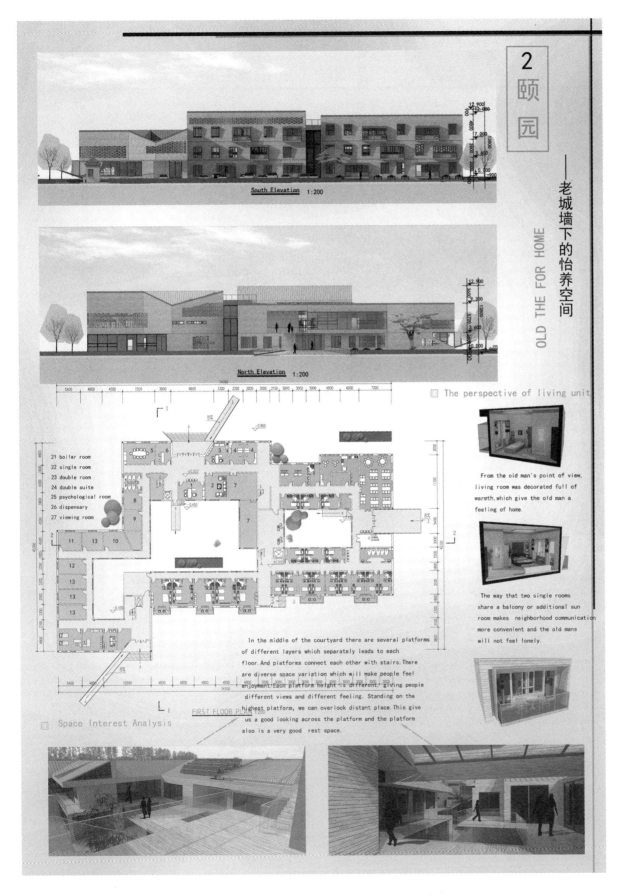

4 颐园
——老城墙下的怡养空间

HOME FOR THE OLD

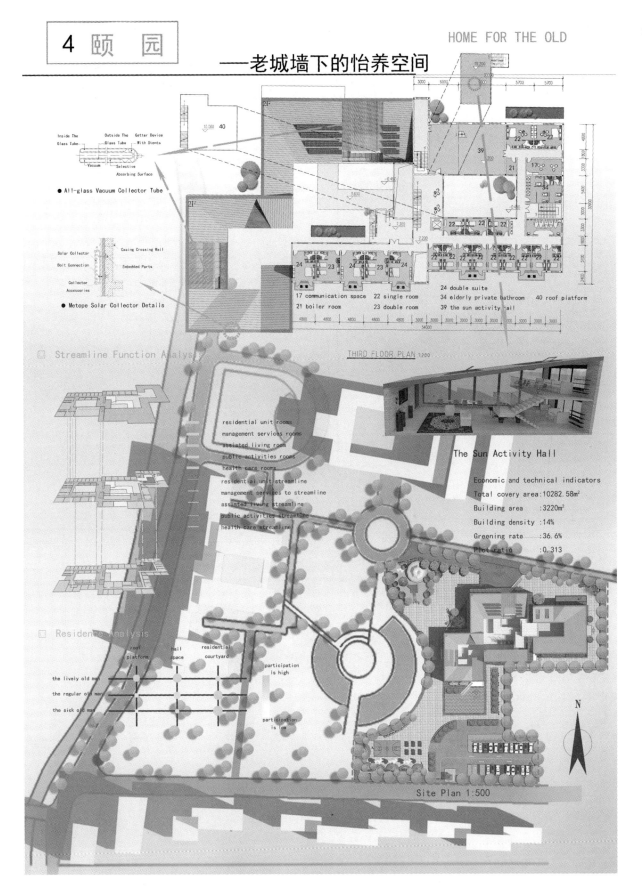

5 颐园

HOME FOR THE OLD

——老城墙下的怡养空间

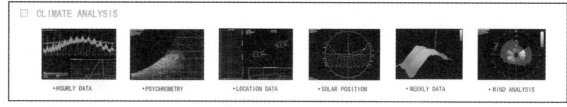

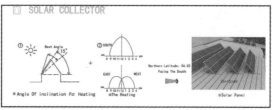

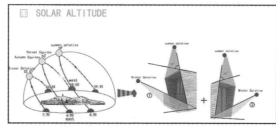

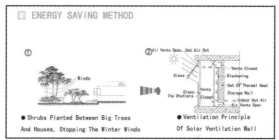

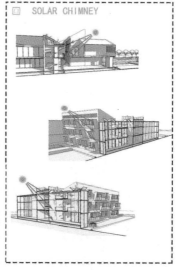

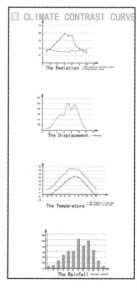

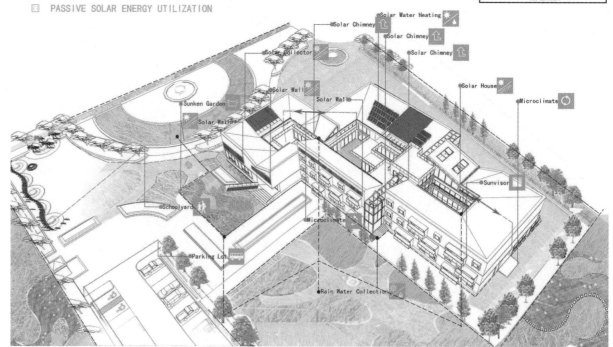

6 颐园
HOME FOR THE OLD
——老城墙下的怡养空间

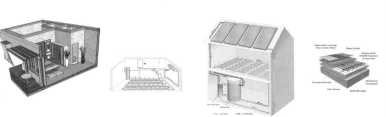

We set a central heating system in living room. It's low temperature hot water floor radiant heating

pension service facilities

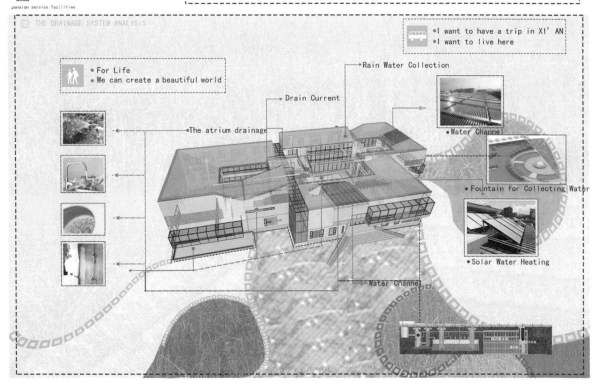

THE DRAINAGE SYSTEM ANALYSIS

- I want to have a trip in XI'AN
- I want to live here

- For Life
- We can create a beautiful world

- Rain Water Collection
- Drain Current
- The atrium drainage
- Water Channel
- Fountain for Collecting Water
- Solar Water Heating
- Water Channel

藤·廊 VENGEVINE ARCADE I

综合奖·优秀奖
General Prize Awarded · Honorable Mention Prize

注 册 号：4645
项目名称：藤·廊（西安）
　　　　　Vengevine Arcade（Xi'an）
作　　者：许彦韬、杨 影、文瑞琳、
　　　　　张艺冰、白洁媛、张清亮
参赛单位：石家庄铁道大学、西安建筑科
　　　　　技大学、中铁建安工程设计院
指导老师：高力强、何国青

藤蔓：以藤蔓象征生命的生生不息，通过人文关怀回应养老中心的定位
绿廊：以生态绿廊延续城市的绿轴，通过绿色体验向公众传达绿色理念

西安生态颐养服务中心选址于西安世界地质公园秦岭山脉终南山下子午镇台沟村。方案致力于探索养老中心与城市绿轴的联系，尝试通过一条体验绿色的生态廊道延续绿轴。以藤蔓的象征意义达到人文关怀的目的。创造一个绿色颐养的活力空间。

Vengevine Arcade

Xi'an ecological service center is located in World Geological Park in Qinling Zhongnanshan mountain ZiwuTown Taigou village. The program is dedicated to exploring pension center and the urban green axis and links to try Green ecological corridor to continue green axis. The symbolic means of humanistic care tovines. Creating a green home activity space.

场地区位分析 / Site Analysis

体块生成分析 / Site Analysis

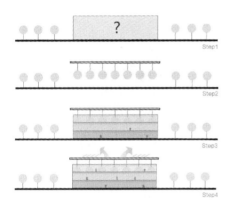

step1　思考使用何种方式处理建筑、城市绿轴及养老空间之间的联系
step2　采用绿化悬挂的设计策略，以生态屋面的形式延续规划绿轴
step3　引入人文关怀的藤蔓形式，以藤蔓象征着老年人也拥有生机
step4　结合场地周边的绿化设计，以交流体验的模式传达绿色概念

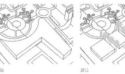
开始 / Initiation　　避让 / Avoid　　穿廊 / Traboules　　庭院 / Courtyard

联通 / Link　　藤廊穿插 / Corridor Alternate　　太阳能X竖悬挂系统 / Solar X Hanging Plant　　立面细化 / Facade Detailed

This design pays attention the ways in which the relationship between architecture, urban green axis, and endowment space is considered. The design strategy of green suspension is used to extend the green axis in the form of ecological roof. The introduction of humanistic care to symbolize the vine form the elderly also have vitality. Combined with the green design around the site, to communicate the experiences of the model to convey the concept of green.

ARCADE II

经济技术指标:
- 用地面积: 8060.85 m²
- 建筑面积: 3310.58 m²
- 占地面积: 2013.35 m²
- 容积率: 0.37
- 建筑密度: 24.98%
- 绿化率: 68.56%
- 停车位: 15
- 单车车位: 50

总平面图 1:600
site-plan

场地气候分析 / Climate Analysis

西安收年平均气温是13.3℃，一月是全年最冷月，平均气温为0.3℃，七月为全年最热月，平均气温是26.9℃。冬季寒冷干燥，夏季炎热多雨。

从图上看不同朝向太阳辐射的差异，南向、西向、东向的辐射量较东西在秋冬季节有较大的差别。

从焓湿图和最佳朝向分析来看，自然通风是该建筑可以采取的较为合理的策略设计。建筑的最佳朝向为南向。

从风玫瑰图来看，全年以东北风为主。夏季主导风为东南风和西南风，冬季以东北风为主。

从综合评价来看，西安近5至8月气候为湿热，3到5月太阳辐射量最大，东向的建筑朝向是最结构化，因此通风是最值得利用的。

结论与策略 / Conclusions

新注意事项：1.建筑应避让跟向为南向或东南向、南西向。2.利用内部通风来排使夏季东北风。
主动策略：1.太阳辐射丰富，可以利用太阳能光伏发电。2.组织雨水湿地集中进行雨水处理。3.充分利用地热能。

Passive strategy: 1. The building best move to the South or South East, South West.
2. The use of natural ventilation and the northeast wind in summer.

Active strategy: 1. Solar radiation is rich using solar photovoltaic power generation.
2. Organization of rainwater collection wetland for water treatment.
3. Rational and rational use of geothermal energy.

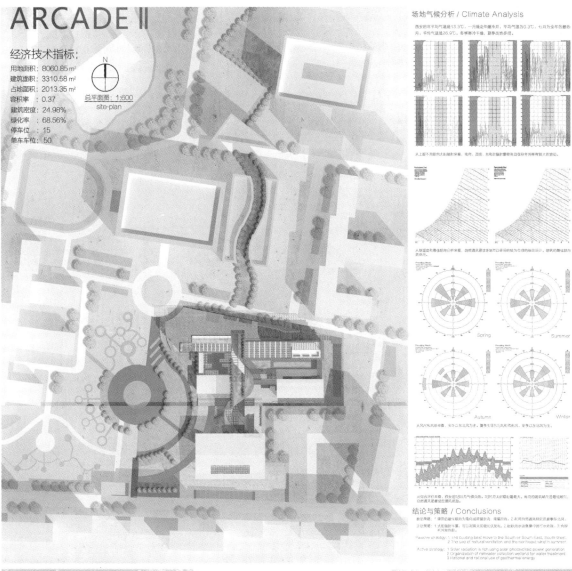

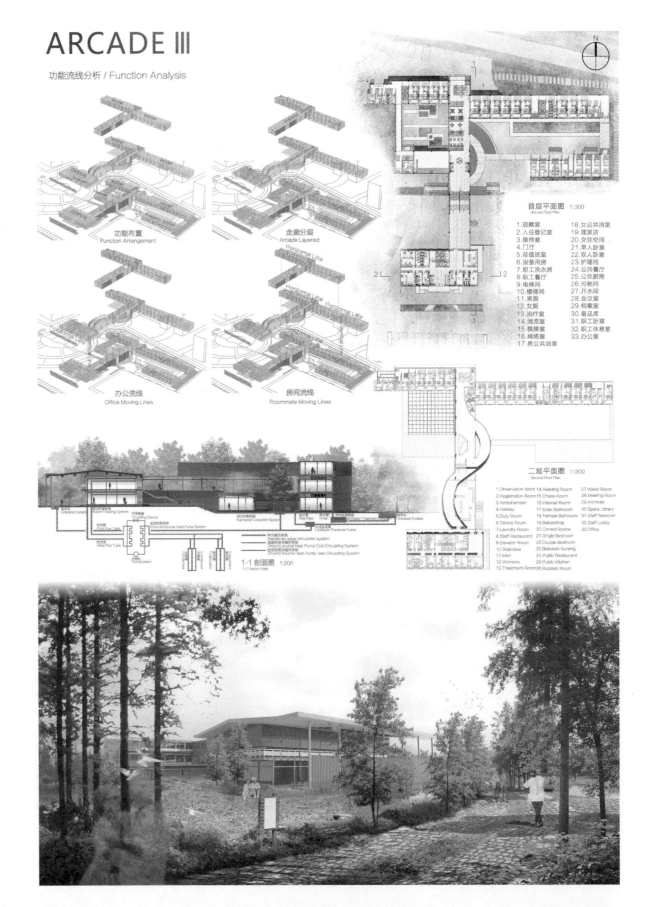

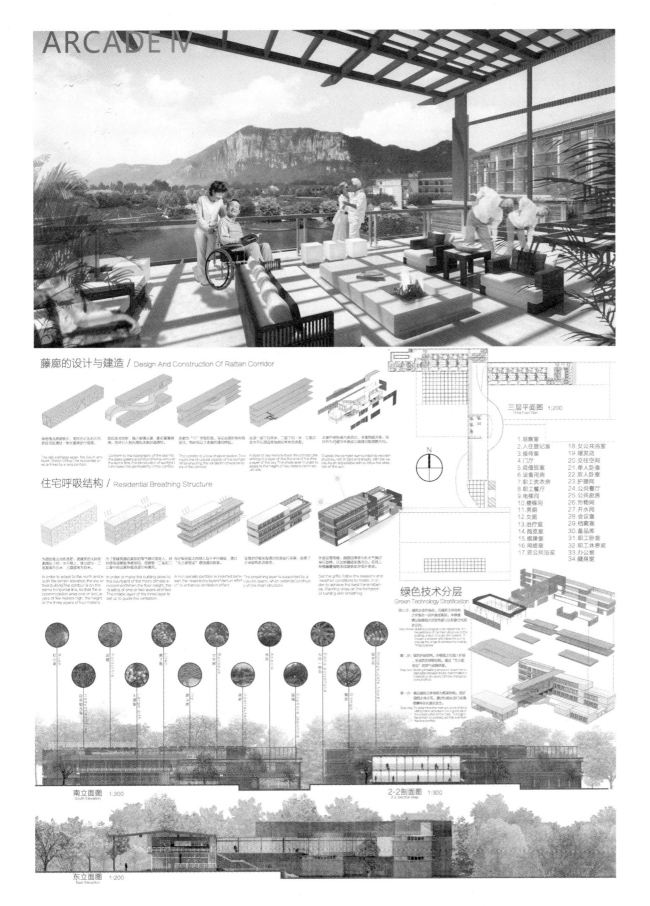

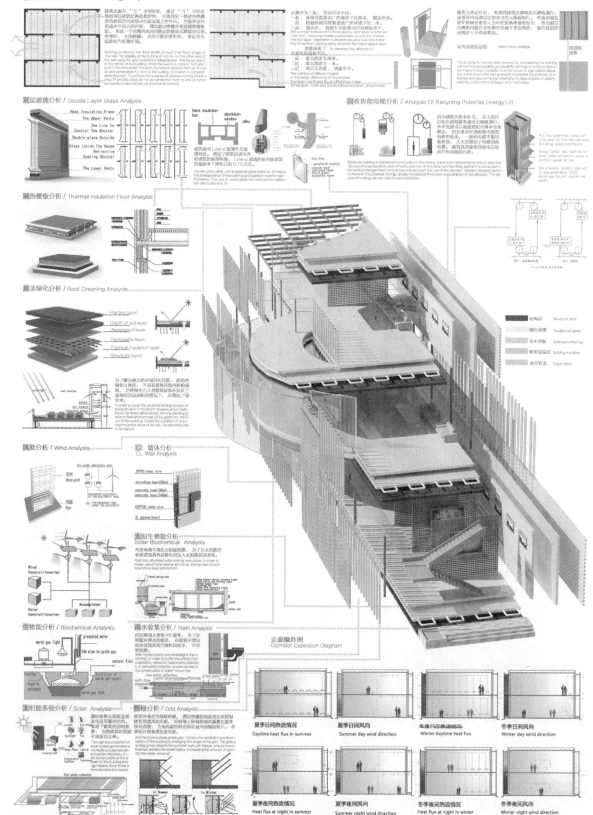

ARCADE VI

太阳能专项分析 / Solar Energy Special Analysis

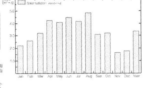
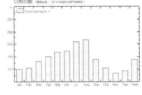
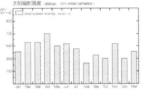
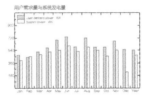

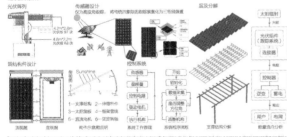

风能专项分析 / Wind Energy Special Analysis

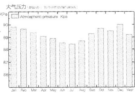
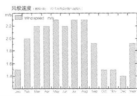

综合奖·优秀奖
General Prize Awarded · Honorable Mention Prize

注 册 号：4683
项目名称：观山居（西安）
　　　　　Mountain House（Xi'an）
作　　者：杨丽、张丹、王泽、
　　　　　高子月、王艳柯、吴嘉俐
参赛单位：西安科技大学
指导老师：孙倩倩

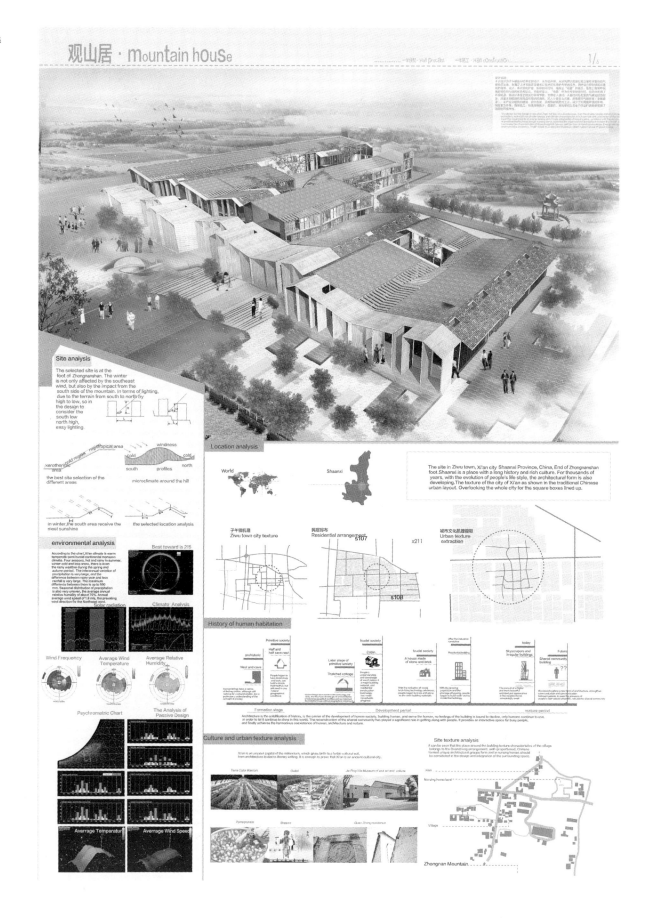

观山居 · mountain house

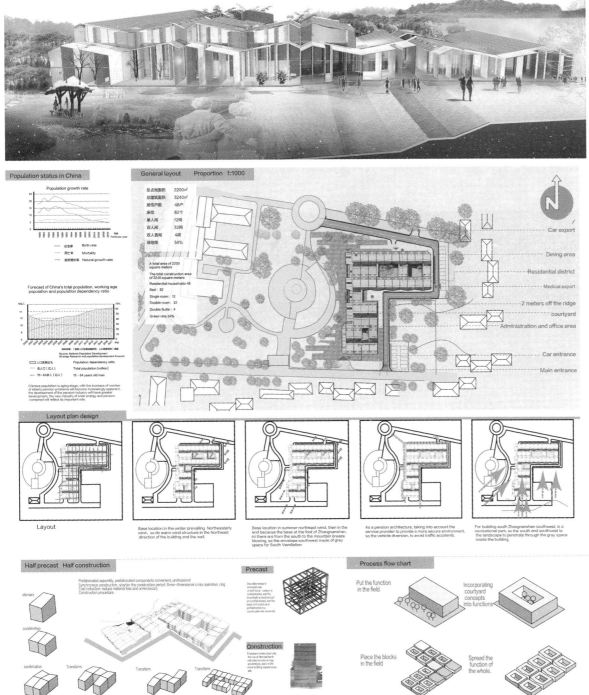
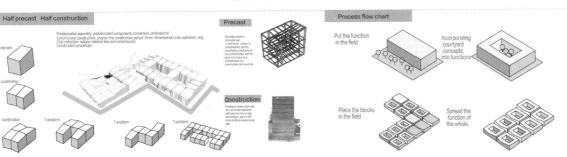

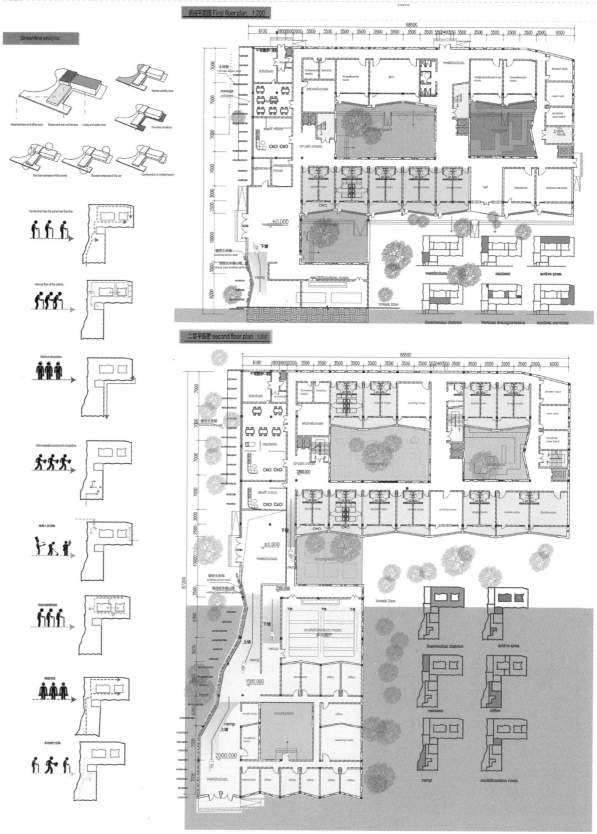

观山居 · mountain house

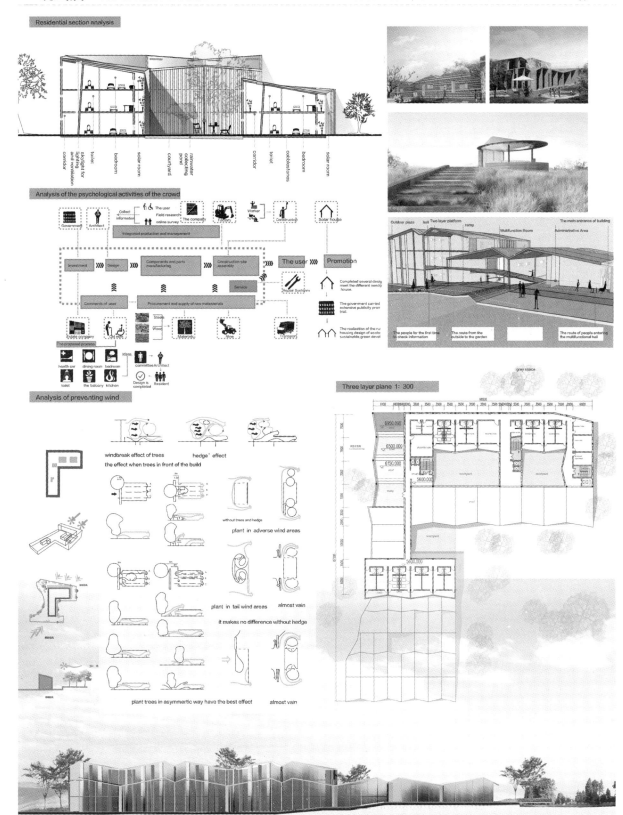

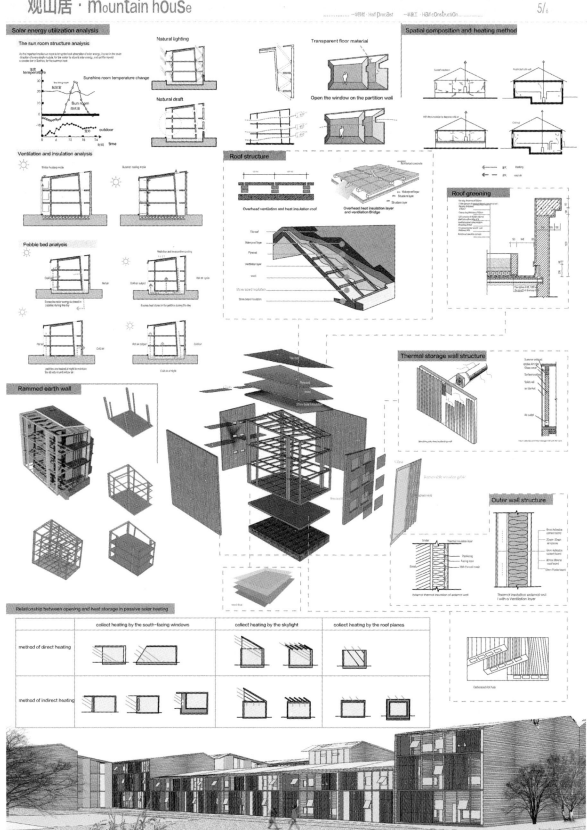

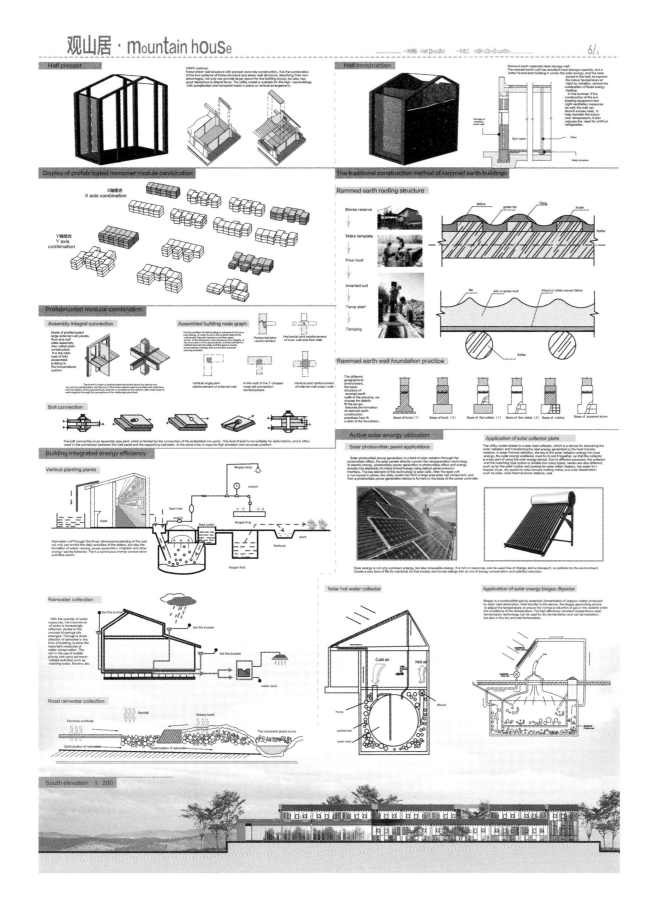

阳光花园 生态颐养服务中心
SUNSHINE GARDEN ECOLOGICAL REMAINING SERVICE CENTER 01

综合奖・优秀奖
General Prize Awarded・
Honorable Mention Prize

注 册 号：4690
项目名称：阳光花园 生态颐养服务中心（西安）
Sunshine Garden Ecological Remaining Service Center (Xi'an)
作 者：刘宜鑫、蒲宏宇、汪涟涟、汪漪漪、唐伟豪
参赛单位：西南科技大学、苏州大学
指导老师：高 明、蔡余萍、成 斌

Site Location Analysis
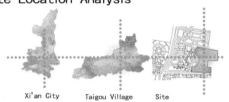
Xi'an City　　Taigou Village　　Site

Traditional Architectural

Tibeton color　　Ornament　　Courtyard　　Eaves

Traditional Culture/Life

Dance　　Theatre　　Food　　Calligraphy

Climate Analysis
   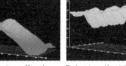
Summer Wind　　Winter Wind　　Average Temperature Yearly　　Relative Humidity Yearly

Design Description
The case on the basis of analysis of the present conditions of the site and the climatic feature of Xi'an, and the residential architectural features of the central Shaanxi plain and habits of local people; solves the problems of heating in winter and ventilation in summer with the ventilation towers, solar house, chasing light blinds, energy-caving insulation wall; and finally determines the placement of solar hot water and solar photovoltaic on the basis of the analysis of the light environment simulation.

设计说明
本方案在充分分析场地现状、西安的气候特点、关中住宅的建筑特色以及当地人民的生活习惯的基础上，以三墙、三院、三楼为概念糅合了墙与院的空间，形成了建筑形体。通过楼梯的通风井、阳光房、遮光百叶、节能保温墙体，充分利用当地日照时间长、太阳辐射强的优势，解决冬季集热采暖和夏季通风降温的问题。最后，基于光环境分析，确定太阳能集热器和太阳能光伏发热安放。
本疗养院融入互联网＋和花卉种植的概念，使老年人生活多元化、兴趣多元化。

Dedign Strategy
1. The use of solar energy, biomass and other renewable energy sources.
2. Passive energy-saving technology mainly, considering positive technology.
3. Local material and conventional construction method.

The Space Requirement of Old Man

Cognitive　Exchange　Recreation　Innovate　Rest　Creation　Contact　Privacy　Culture　Personal

阳光花园 生态颐养服务中心
SUNSHINE GARDEN ECOLOGICAL REMAINING SERVICE CENTER 02

Data Investigation

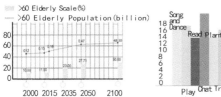

Site Analysis

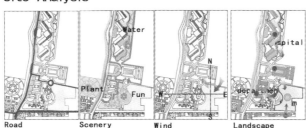

Road | Scenery | Wind | Landscape

Concpt Generation

Separation of architecture and environment → Integration into the Environment / The natural space shuttle in flowers

Lack of communication space → Create garden, exchange space / Garden type layout, nodes form a centripetal exchange place

Lack of emotional and memory of the living space → Place spirit / Guanzhong folk house, vertical greening, flower planting

Idle Lonely — Action inconvenience, unable to go out — Low sense of belonging, Life is dull → Play in the garden, studing different gardening techniques — High sense of belonging, Interesting life — Vertical greening

Internet Plus Pension

Fill ill / Medical treatment / Health data cloud upload / Fast Efficient Safe / Chatting on the Internet / Online appointment for medical examination / Intelligent Call / Warm and comfortable

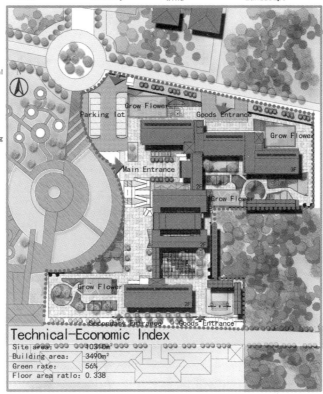

Technical-Economic Index
Site area: 10310m²
Building area: 3490m²
Green rate: 56%
Floor area ratio: 0.338

SITE-PLAN 1:500

Scheme Generation

Linking Axis Generation | Public Space Analysis
Flow Analysis | Extension Analysis

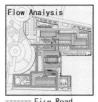 

Traffic / Park / Square / Grow flower

Fire Road / Staff's Car Flow / Pedstrain Flow / Second Stage Construction / First Stage Construction

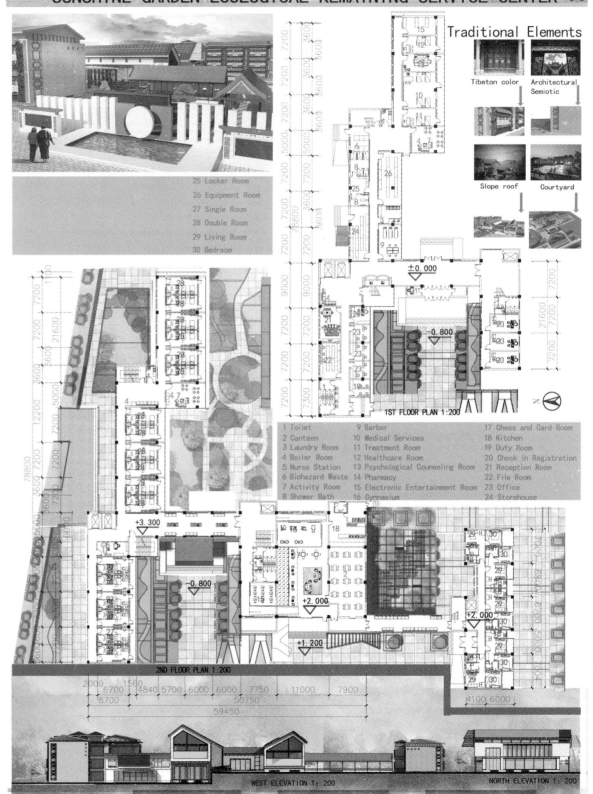

阳光花园 生态颐养服务中心
SUNSHINE GARDEN ECOLOGICAL REMAINING SERVICE CENTER 04

7 Deluxe Suite A
8 Deluxe Suite B
9 Deluxe Suite C
10 Deluxe Suite D
11 Deluxe Suite E

Double-Glazed

Winter: Hot air through the window up and down into the indoor. Play the role of insulation

Gale: Close the outer glass window to keep the wind. The intermediate layer plays the role of air circulation

Summer: To open the outer glass window, guide the wind into the room. Layer on to let the wind into the room

Surface Rainwater Collection System　Microclimate

Taking in water — Pretreatment — Aquatic vegetation — Recycling use
Matrix, plant and microbial complex
The waste water is filtered through the sewage and used by people
Trees absorb carbon dioxide and other gases

Ecotect Data Analysis

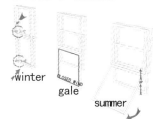

Analysis of winter solstice
Analysis of incident radiation in winter

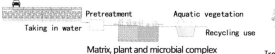

Sunshine shadow analysis
Analysis of winter solstice lighting
spring　summer　autumn　winter

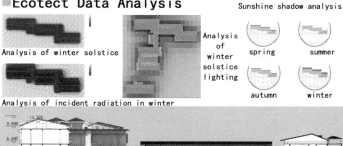

3rd FLOOR PLAN 1:200

1 Toilet
2 Boiler room
3 Activity room
4 Nurse station
5 Double room
6 Single room

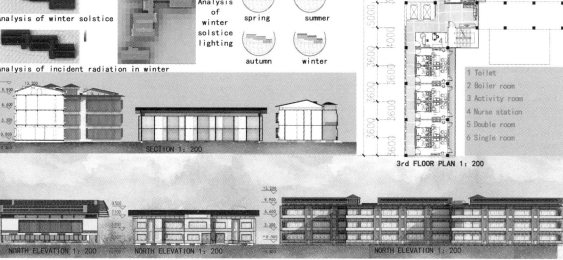

SECTION 1:200　　NORTH ELEVATION 1:200　　NORTH ELEVATION 1:200　　NORTH ELEVATION 1:200

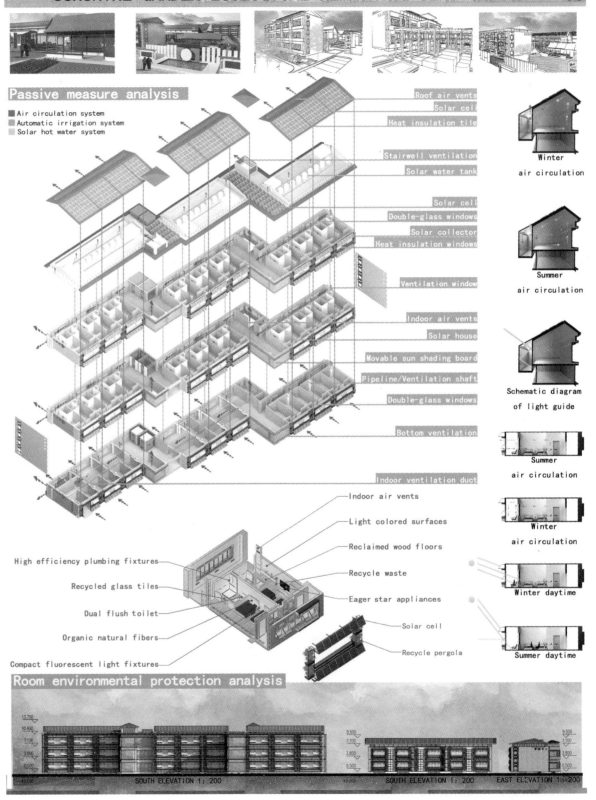

阳光花园 生态颐养服务中心
SUNSHINE GARDEN ECOLOGICAL REMAINING SERVICE CENTER 06

Detail 1:
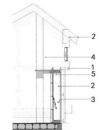

1. Photovoltaic elements
2. Radiant ceiling
3. Double glazed back enameled panel
4. Variable solar panel regulating system
5. Structure column
6. I-beam
7. Ceiling troch disc

Detail 2:

1. Exterior shade panel
2. Operable window
3. Steel bracket
4. Exposed concrete
5. Fire insulation infill

Rainwater Collection Storage and Re-use

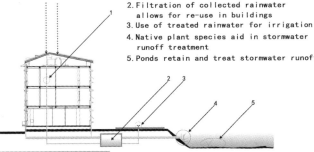

1. Re-use rainwater in toilets
2. Filtration of collected rainwater allows for re-use in buildings
3. Use of treated rainwater for irrigation
4. Native plant species aid in stormwater runoff treatment
5. Ponds retain and treat stormwater runoff

Detail 3:
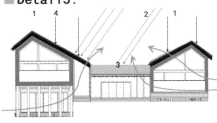

1. Slanted roof is installed with solar photovoltaic panels
2. Skylight can protect natural lighting in daytime and avoid direct solar radiation in summer
3. Extensive openings and skylights provide sufficient convection current in summer
4. The solar house heats the surrounding room providing adequate lighting

Facade Elements

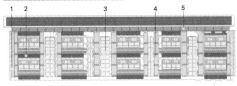

1. Vent between the two walls
2. Solar panels
3. Glass screen wall
4. Plants planted on the facade
5. Lighting curtain wall

Energy Analysis

$1155 \times$ ☒ $\times 365 \text{(days)} = 161,700 \text{(Kilowatt hour)}$

$= 400 \times$ ☒ $\times 421 \text{(days)}$

$= 100 \times$ ☒ $\times 1080 \text{(hours)}$

$= 30 \times$ ☒ $\times 2156 \text{(hours)}$

$= 20 \times$ ☒ $\times 16170 \text{(hours)}$

$= 100 \times$ ☒ $\times 8085 \text{(hours)}$

$= 80850 ¥ \text{ per year}$

$2200 m^2 \times$ ☘ $= +110kg \ O_2 - 130kg \ CO_2 \text{ (per day)}$

$= 150 \times$ $\times 1 \text{ (days)}$

The Sunshine Condition
Winter solstice

6:00 am

10:00 am

4:00 pm   6:30 pm

综合奖・优秀奖
General Prize Awarded・
Honorable Mention Prize

注 册 号：4720
项目名称：光院・居（西安）
　　　　　Sunshine and Yard Build the House（Xi'an）
作　　者：张云卿、胡庆强、董毓兵
参赛单位：石家庄铁道大学
指导老师：高力强、何国青

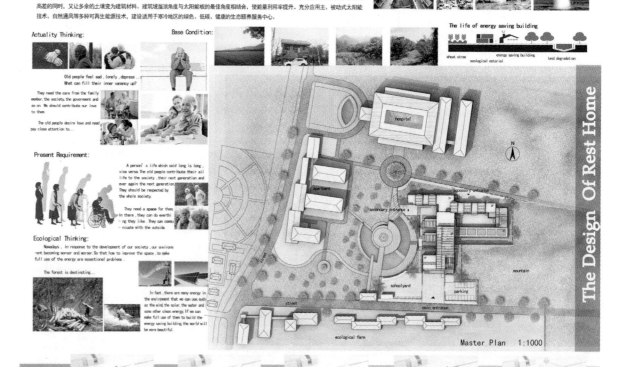

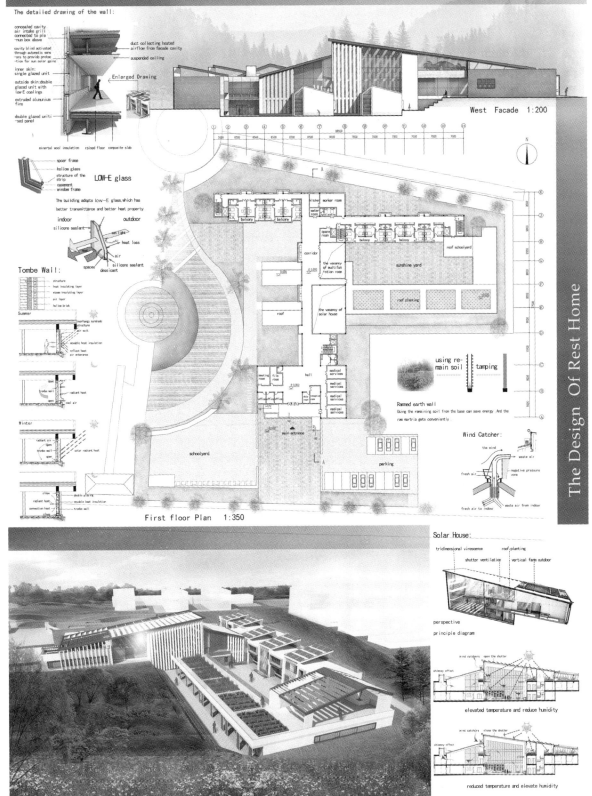

The Design Of Rest Home

Sunshine And Yard Build The House 3 光院·居

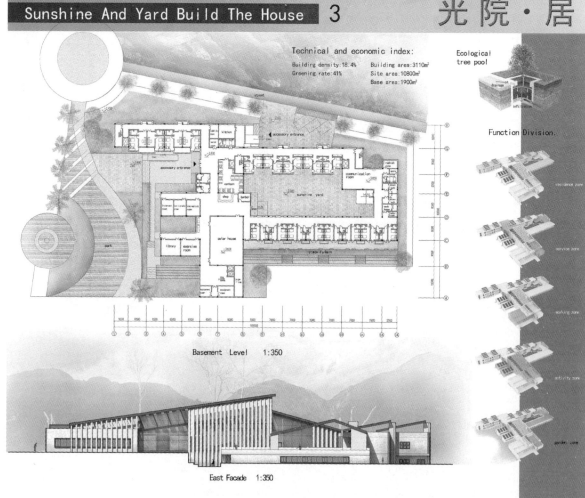

Basement Level 1:350

East Facade 1:350

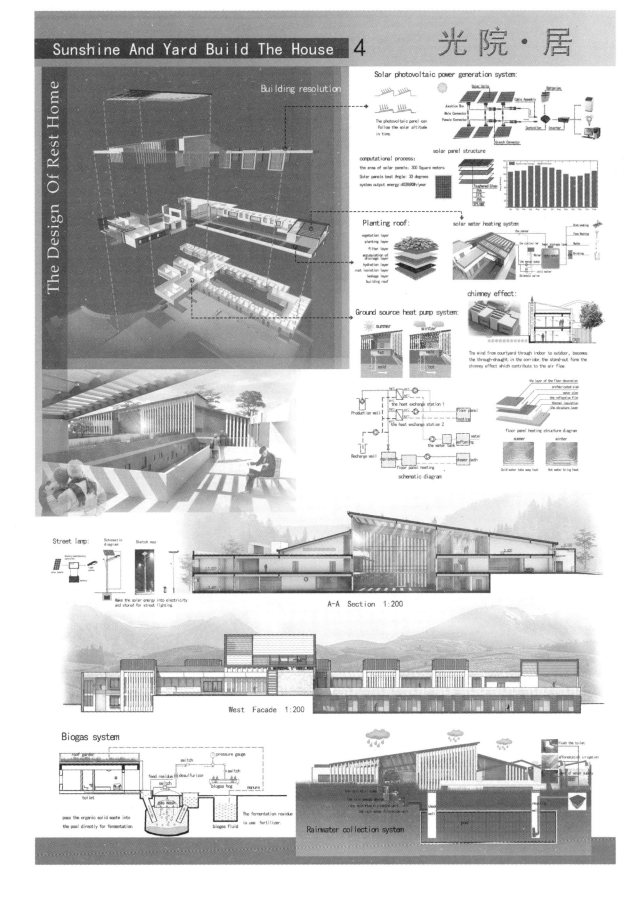

综合奖·优秀奖
General Prize Awarded · Honorable Mention Prize

注册号：4749
项目名称：暖院·悟境（西安）
　　　　　Solar & Silent Yard（Xi'an）
作　者：刘畅、李重锐、吴琼、李庆祥
参赛单位：天津大学
指导老师：朱丽、严建伟

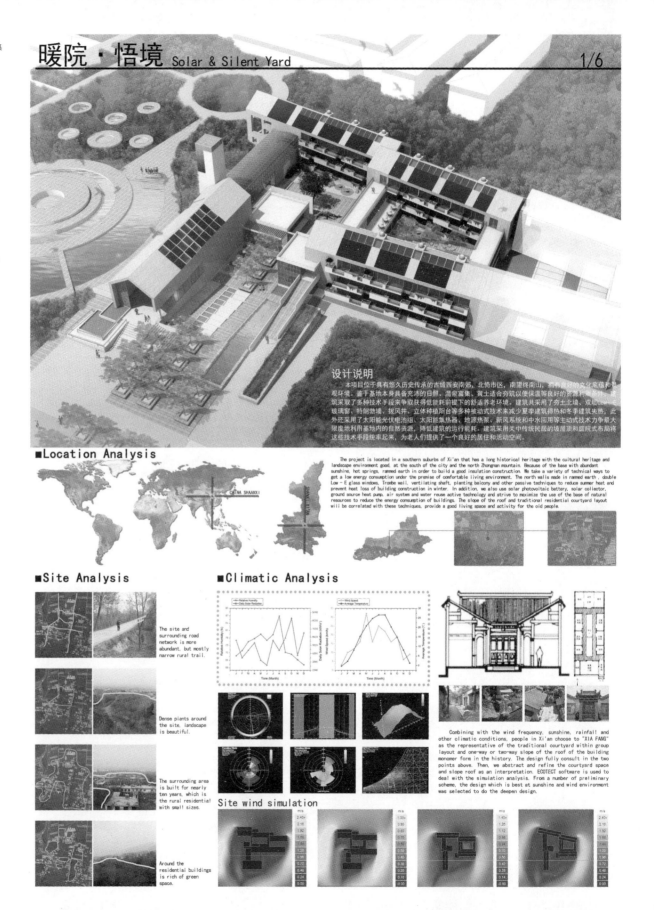

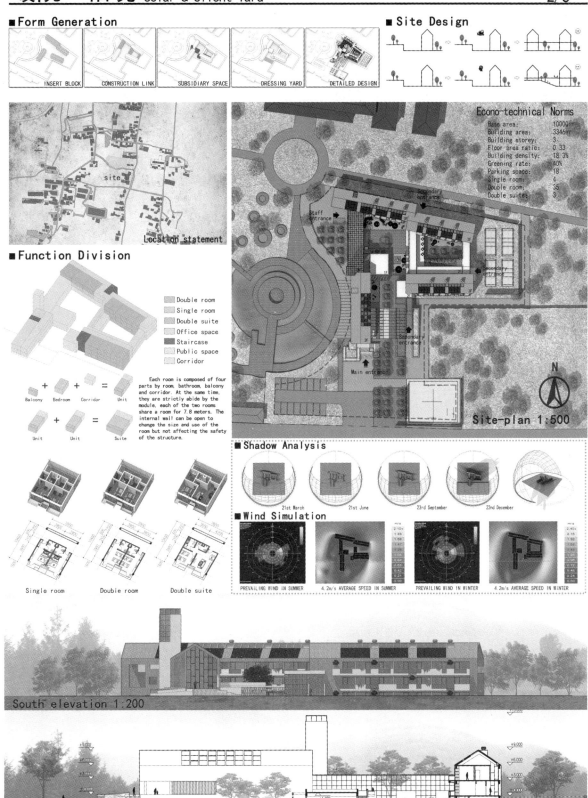

暖院·悟境 Solar & Silent Yard

■ Pension Service Status

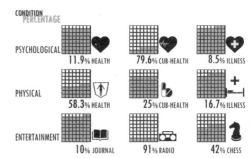

With the China aging trend is increasing, more and more elderly people need to enter nursing homes and other service institutions, but such institutions in China to provide the service is still at the low level, the old people are lack of venues.

The number of people living in nursing homes exceeded the design capacity, and the building itself is not reserved for expansion, which lead to a room into the far more than the number of the design, living space is crowded and messy.

At present, many of the old building is not designed for the pension function, but to take the old building transformation methods meet their requirements. The form of a building is single and with poor environment.

Because of the lack of enough places and activities, the old people have nothing to do with them, which seriously affects their state of mind. In this project, the above problems will be considered and proposed solved.

According to the data provided by the WHO, China's elderly mental health status is not optimistic, 88.1% of the people in the disease and sub-health status, physical health data slightly better: about 58.3% of the elderly health status is good, at the same time, the form of Chinese elder people's entertainment is also single, with radio and chess as a basic entertainment project.

PSYCHOLOGICAL: 11.9% HEALTH | 79.6% CUB-HEALTH | 8.5% ILLNESS
PHYSICAL: 58.3% HEALTH | 25% CUB-HEALTH | 16.7% ILLNESS
ENTERTAINMENT: 10% JOURNAL | 91% RADIO | 42% CHESS

■ New Lifestyle

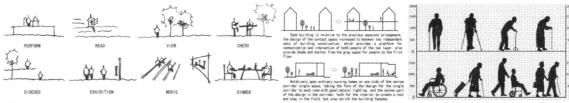

PERFORM | READ | VIEW | CHESS
DISCUSS | EXHIBITION | MOVIE | DINNER

Each building is relative to the previous separate arrangement, the design of the contact space increased to between two independent sets of building construction, which provides a platform for communication and interaction of both people of the two layer, also provide shade and shelter from the gray space for people to the first floor.

Relatively open ordinary nursing homes on one side of the narrow corridor single space, taking the form of the design for the single corridor to each room with good natural lighting, and the convex part of the design in the corridor, both for the interior to create a rest and stay in the field, but also enrich the building facades.

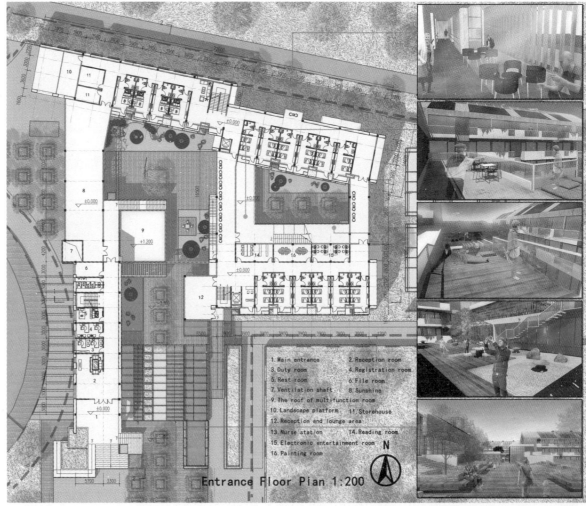

1. Main entrance
2. Reception room
3. Duty room
4. Registration room
5. Rest room
6. File room
7. Ventilation shaft
8. Sunshine
9. The roof of multifunction room
10. Landscape platform
11. Storehouse
12. Reception and lounge area
13. Nurse station
14. Reading room
15. Electronic entertainment room
16. Painting room

Entrance Floor Plan 1:200

暖院·悟境 Solar & Silent Yard

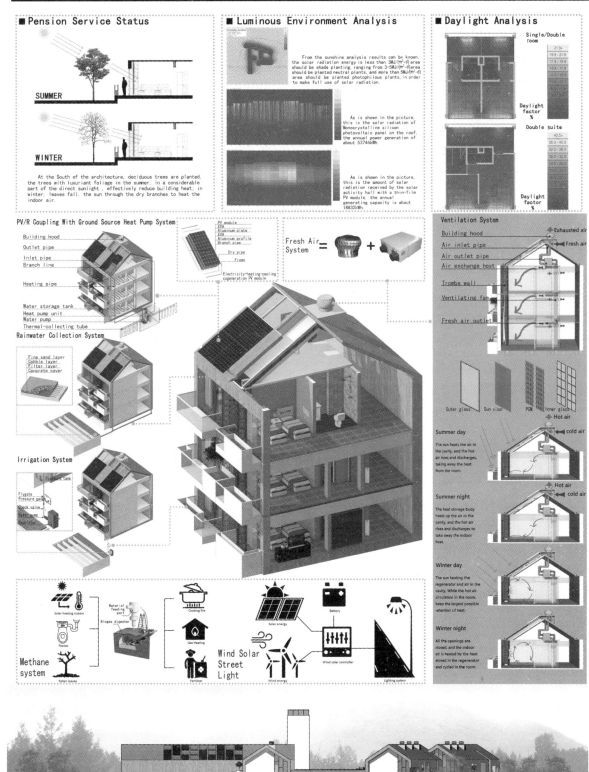

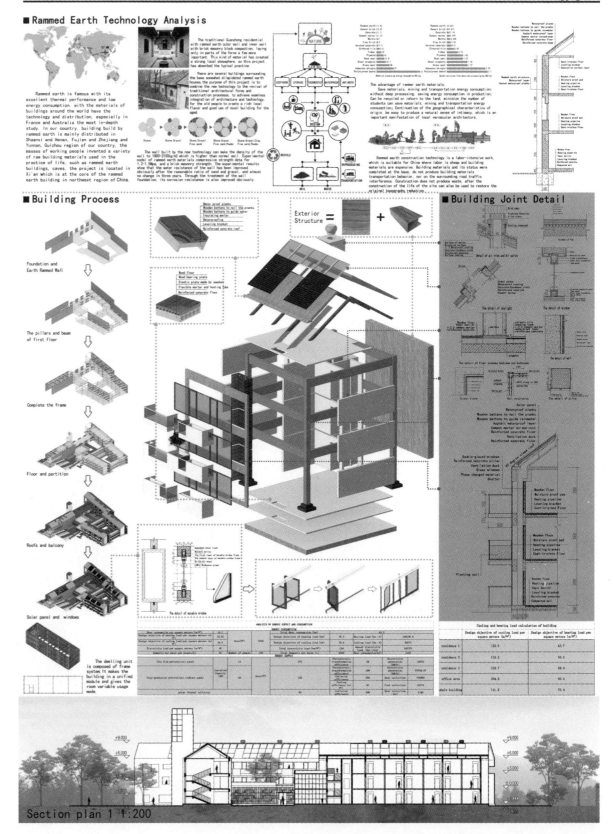

暖院·悟境 Solar & Silent Yard

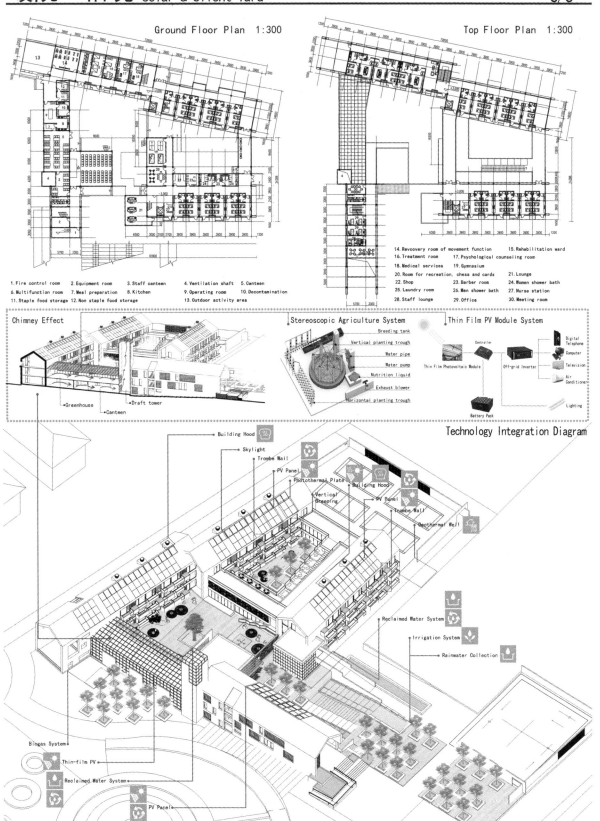

综合奖·优秀奖
General Prize Awarded · Honorable Mention Prize

注 册 号：4830
项目名称：沐光之城（西安）
　　　　　Via Light（Xi'an）
作　　者：黄　华、郑智中、李　真、
　　　　　赵　旭、徐洪光、薛芳慧、
　　　　　刘增军、刘　冲
参赛单位：西安建筑科技大学
指导老师：李　钰、罗智星

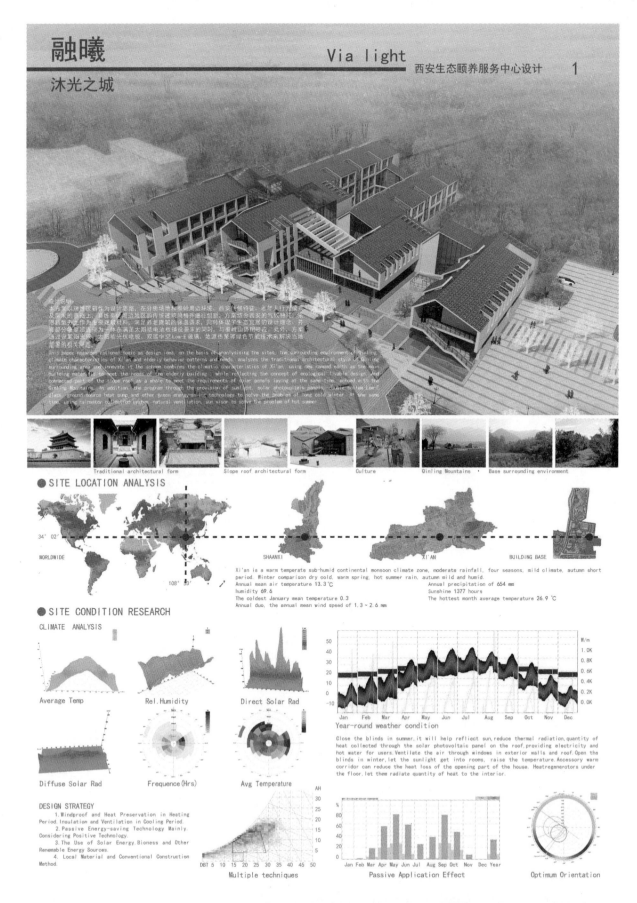

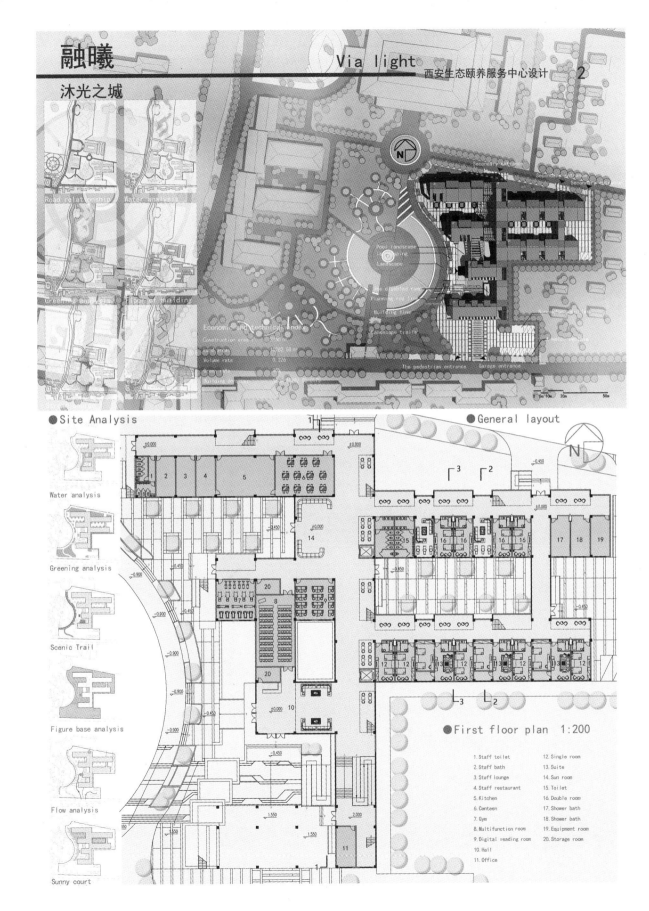

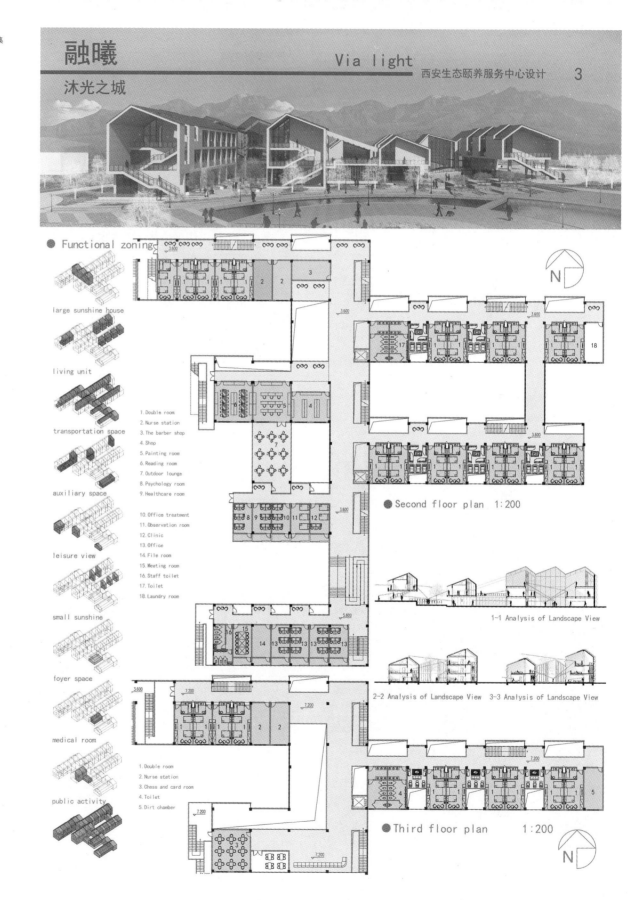

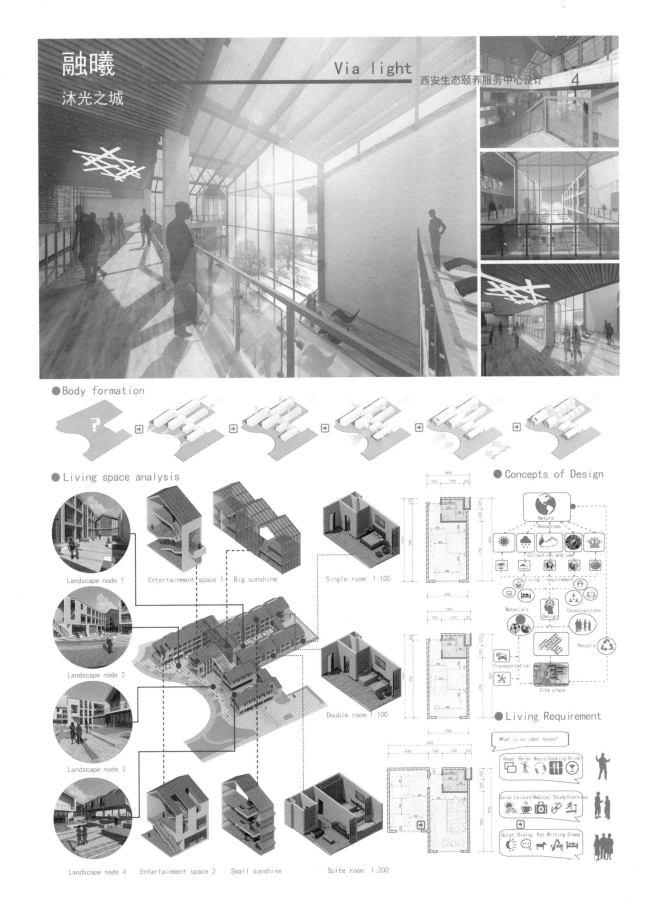

融曦 Via light
沐光之城
西安生态颐养服务中心设计

● 1-1 perspective view

● Solar hot water system
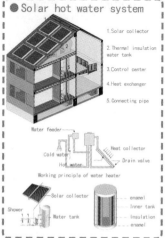
1. Solar collector
2. Thermal insulation water tank
3. Control center
4. Heat exchanger
5. Connecting pipe

● Rainwater collection
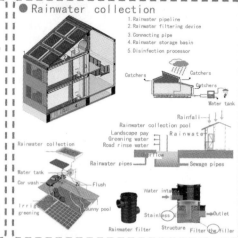
1. Rainwater pipeline
2. Rainwater filtering device
3. Connecting pipe
4. Rainwater storage basin
5. Disinfection processor

● Gshps
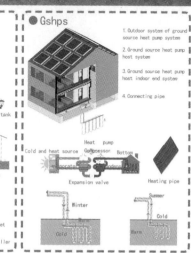
1. Outdoor system of ground source heat pump system
2. Ground source heat pump host system
3. Ground source heat pump host indoor and system
4. Connecting pipe

● Composite wall
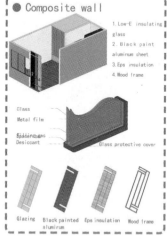
1. Low-E insulating glass
2. Black paint aluminum sheet
3. Eps insulation
4. Wood frame

● Architectural analysis
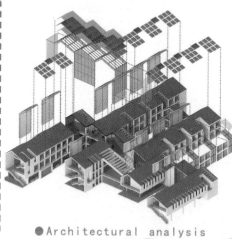

● Passive sunshine room system
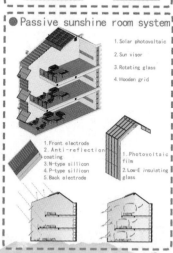
1. Solar photovoltaic
2. Sun visor
3. Rotating glass
4. Wooden grid

● South elevation 1:200

融曦 Via light

沐光之城

西安生态颐养服务中心设计 6

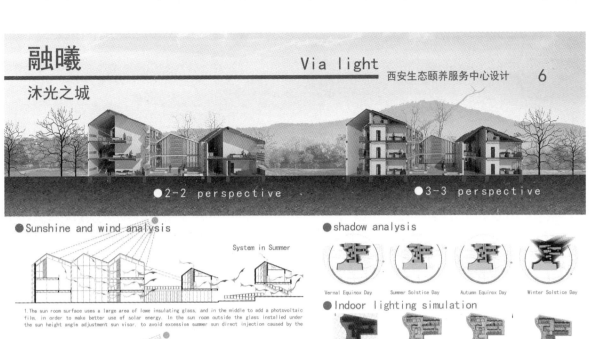

● 2-2 perspective   ● 3-3 perspective

● Sunshine and wind analysis

1. The sun room surface uses a large area of lowe insulating glass, and in the middle to add a photovoltaic film, in order to make better use of solar energy. In the sun room outside the glass installed under the sun height angle adjustment sun visor, to avoid excessive summer sun direct injection caused by the

2. In order to prevent the impact of the winter monsoon, the sun room on the north side did not use a large area of glass, but with a thicker rammed earth wall. This has played a good wind insulation effect, in the middle of the sun room to use both sides of the desktop back through the space, is conducive to the air within the loop effect.

Dwelling unit in Summer   Dwelling unit in Winter

3. The external walls of the living unit are constructed in a special form and the thermal loss of the window glass is reduced as much as possible in ensuring adequate lighting. In the vicinity of the corridor side to do a part of the through space to help the room a good air circulation.

4. In order to prevent the impact of the winter monsoon, the living area on the north side of the walls less windows. In order to solve the corridor lighting problems, we opened a part of the roof skylight, combined with through space, so that indoor air circulation is more smooth.

Sunshine room in Summer   Sunshine room in Winter

5. The rest unit uses a glass and sun visor similar to the large sun room, which can effectively avoid excessive sun exposure in summer. Close the blinds in summer, it will help reflect sun, reduce thermal radiation, quantity of heat collected through the solar photovoltaic panel on the roof, providing electricity and hot water for users.

Open the blinds in winter, let the sunlight get into rooms, raise the temperature. Accessory warm corridor can reduce the heat loss of the opening part of the house. Heat regenerators under the floor, let them radiate quantity of heat to the interior.

● shadow analysis

Vernal Equinox Day   Summer Solstice Day   Autumn Equinox Day   Winter Solstice Day

● Indoor lighting simulation

● Structural analysis

1. Tiles
2. Thermal insulation roofing
3. Concrete slab

1. Brick wall
2. Thick layer between air
3. Thickness of polystyrene
4. Fiber reinforced layer

1. Single wood
2. Floor
3. Wooden grid
4. Lime slag
5. Floor
6. The main keel
7. Times keel

1. The surface
2. The structure layer
3. Cushion layer
4. At the grass-roots level
5. Element of soil compaction

● Comprehensive technical analysis chart

Circuit diagram
Ground source heat pump
Solar water heating system
Rainwater collection system

Solar heater   Metope   Solar panels   Heating pipeline   Lowe insulating glass   Solar cell

● East elevation 1:200

● West elevation 1:200

综合奖・优秀奖
General Prize Awarded ·
Honorable Mention Prize

注 册 号：4856
项目名称：漫步・时・光（西安）
　　　　　Time・Light（Xi'an）
作　　者：丁辛宇、丁怡文、鲍程远
参赛单位：吉林建筑大学
指导老师：裘　鞠、李雷立

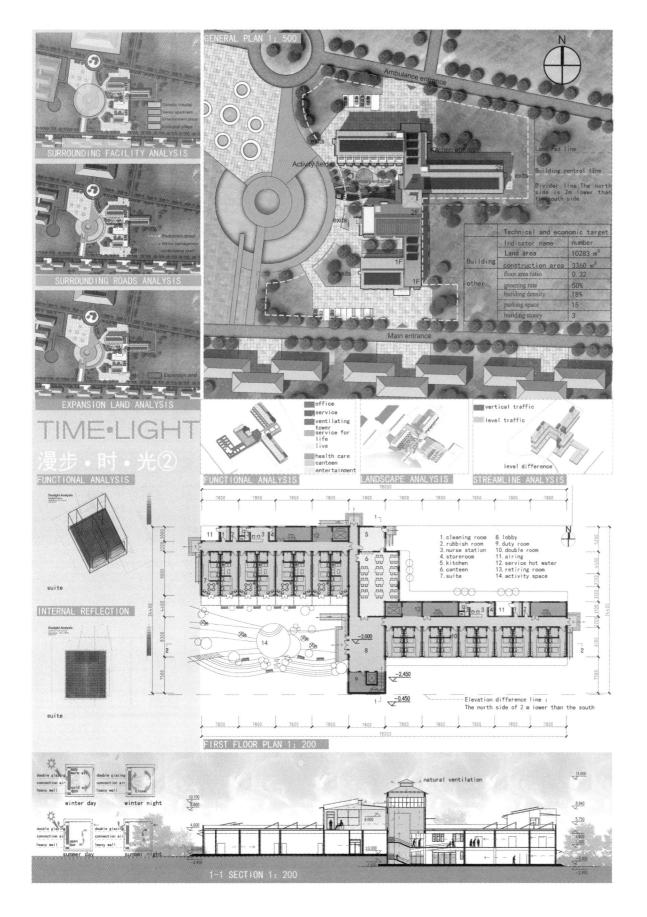

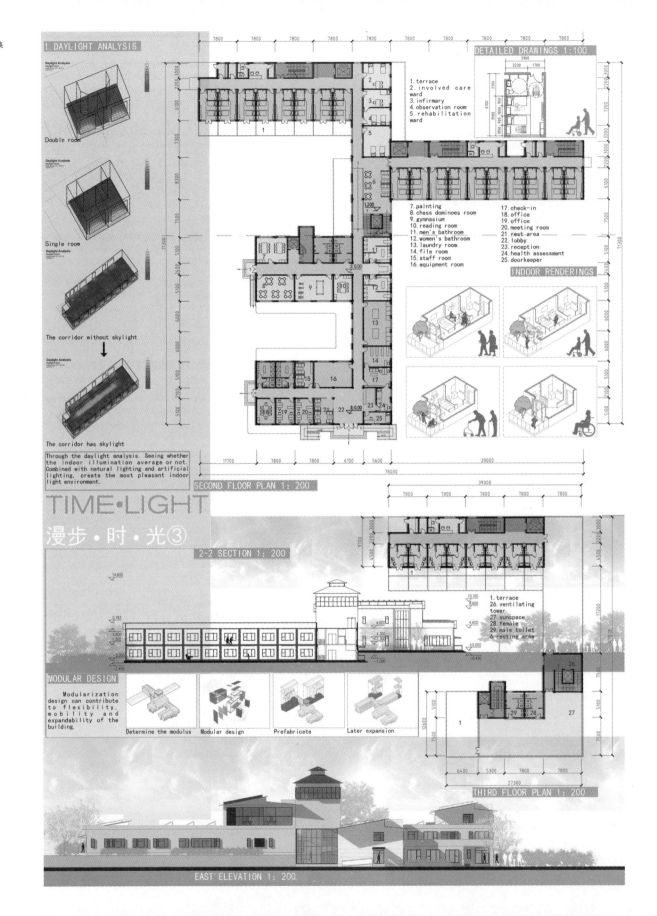

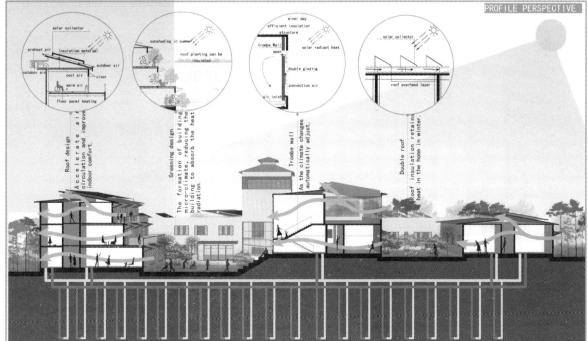

综合奖·优秀奖
General Prize Awarded·
Honorable Mention Prize

注 册 号：4932
项目名称：悠悠然居（西安）
　　　　　Leisurely Living（Xi'an）
作　　者：彭　悦、孟　婕、杨鹏程、
　　　　　朱　硕
参赛单位：南京工业大学
指导老师：胡振宇

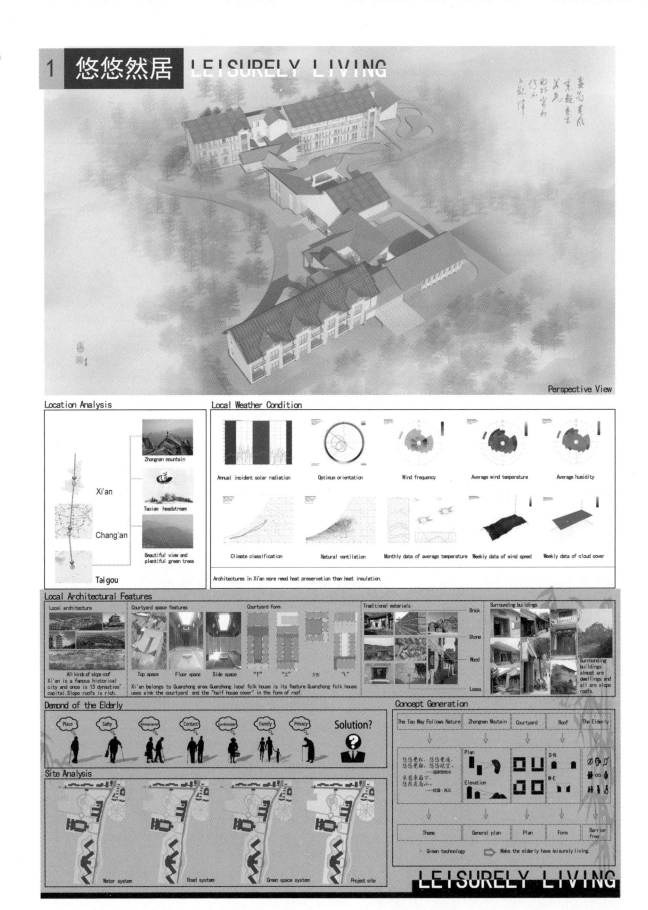

2 悠悠然居 LEISURELY LIVING

设计说明 / Design Description

基地位于西安市长安区子午镇台沟村，南临秦岭终南山，拥有悠久的道教历史，所以本案以"道法自然"思想为指导，确立"悠悠然居"的主题；以西安关中民居的建筑特色为原型，设计庭院式平面和坡屋顶立面造型；功能布局充分围绕老年人生理、心理需求，合理分区，注意无障碍设计；采用合理的平面规划和造型设计，并结合太阳能光电技术、太阳能光热技术、雨水收集等节能措施，旨在建造一个闲适、绿色、适老的养老服务中心。

The site is located at Taigou Village, Ziwu Town, Chang'an District, Xi'an City. On the south side, there is the Qinling Zhongnan mountain. The Taoism history of Zhongnan is centuries-old, so the case follows "the tao way follows nature" and the theme is "leisurely living"; The Guanzhong folk house are used as the prototype of this case, so we design plan with courtyard and elevation form with slope roofs; To meet the elderly's special needs on physiology and psychology, we divide functions reasonably and pay attention to barrier free; We adopt proply general plan design and form design combined with solar photovoltaic technology, solar thermal technology, rainwater collection technology and other saving energy technologies. It is the goal that to build a pension service center which is leisurely, green, suitable for the elderly.

Economic and Technical Norms

Site area	9964.6 m²	Greening rate	48.2%
Building area	3332.7 m²	Foor area ratio	0.33
Gross floor area	1728.6 m²	Parking lots	16
Building density	0.17	Number of beds	80

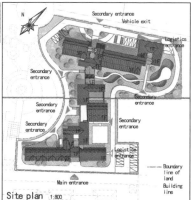

Site plan 1:800

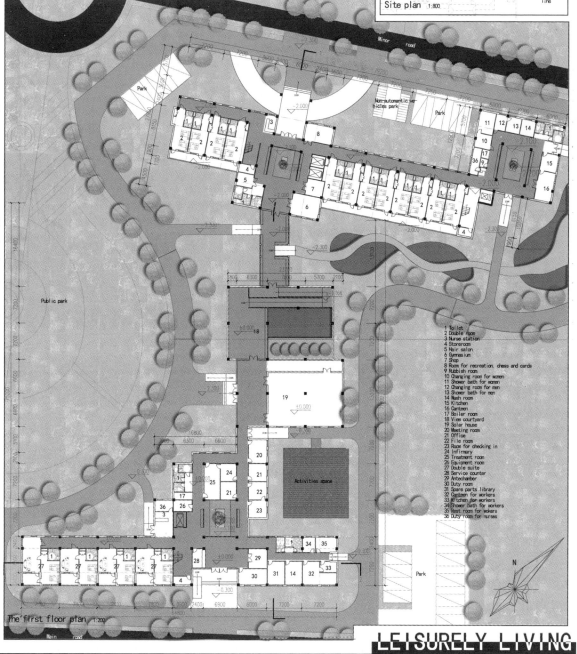

The first floor plan 1:200

1 Toilet
2 Double room
3 Nurse station
4 Storeroom
5 Hair salon
6 Gymnasium
7 Shop
8 Room for recreation, chess and cards
9 Rubbish room
10 Changing room for women
11 Shower bath for women
12 Changing room for men
13 Shower bath for men
14 Wash room
15 Kitchen
16 Canteen
17 Boiler room
18 View courtyard
19 Solar house
20 Meeting room
21 Office
22 File room
23 Room for checking in
24 Infirmary
25 Treatment room
26 Equipment room
27 Double suite
28 Service counter
29 Antechamber
30 Duty room
31 Spare parts library
32 Canteen for workers
33 Kitchen for workers
34 Shower bath for workers
35 Rest room for workers
36 Duty room for nurses

LEISURELY LIVING

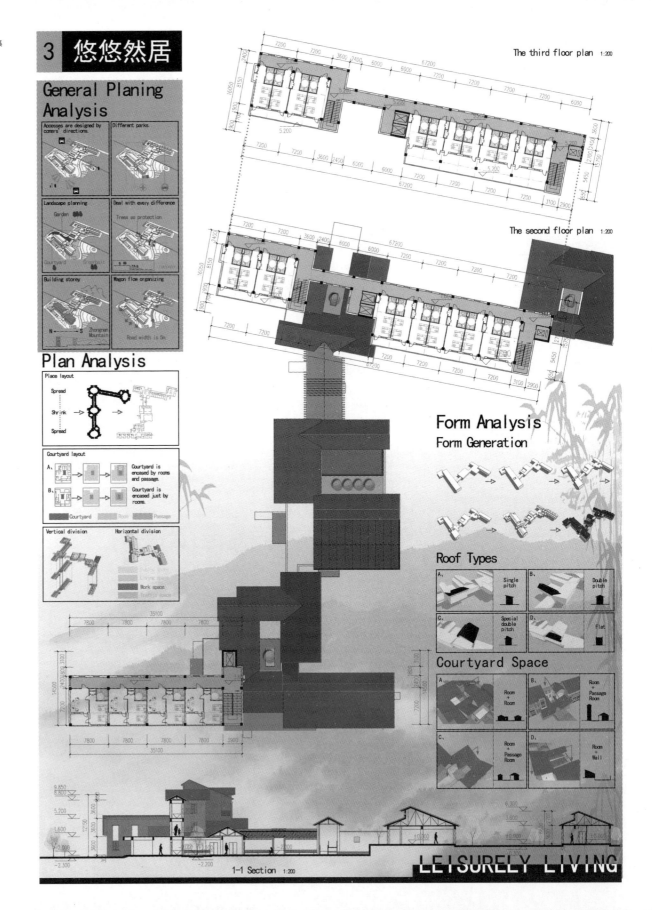

4 悠悠然居
Proper Design For The Elderly

Dwelling Size Analysis

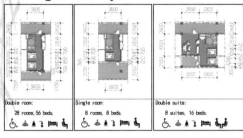

Double room: 28 rooms, 56 beds.
Single room: 8 rooms, 8 beds.
Double suite: 8 suites, 16 beds.

Nursing Uite

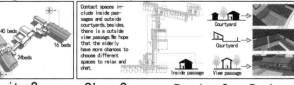

Contact Space

Contact spaces include inside passages and outside courtyards, besides, there is a outside view passage. We hope that the elderly have more chances to choose different spaces to relax and chat.

Activity Space

The elderly could play chess or cards, exercise, dance and so on when they are free.

Sleep Space

It is hard for the elderly to have a good sleep, so we make sleep space surrounded by green belt and trees.

Barrier-free Design

All dwelling rooms' toilets meet the disabled elderly's requirements.

Architecture's entrances all use ramp.

Set up public toilet which is used by the disabled only.

Park for the disabled

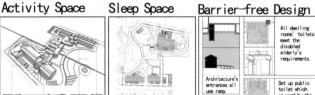

2-2 Section 1:200

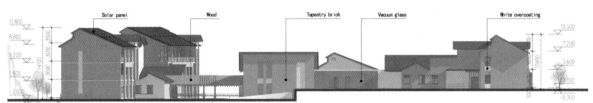

West elevation 1:200

South elevation 1:200

LEISURELY LIVING

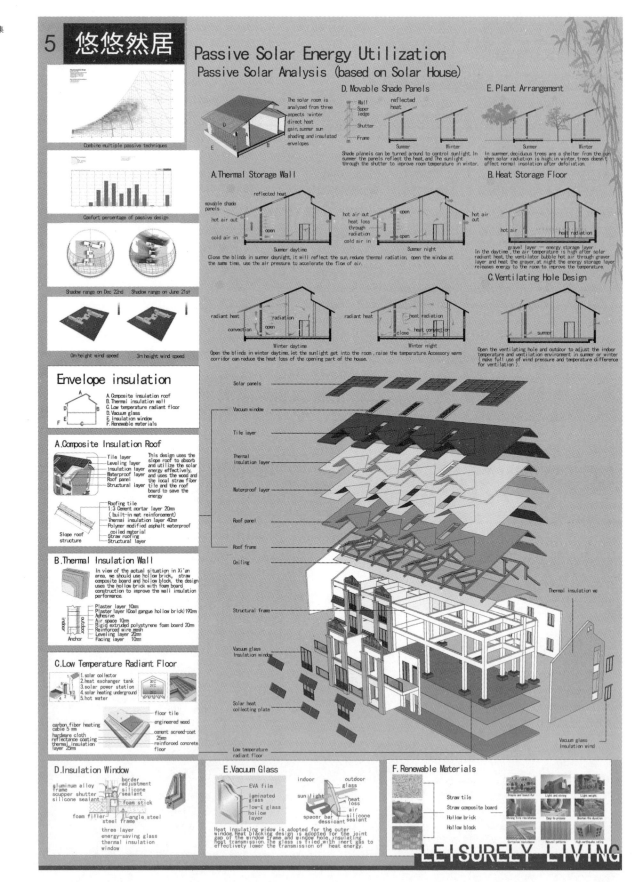

6 悠悠然居 Active Solar Energy Utilization And Other Technologies

A. Solar Water Heating System

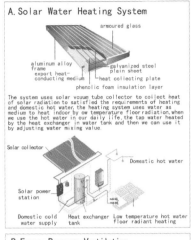

The system uses solar vacuum tube collector to collect heat of solar radiation to satisfied the requirements of heating and domestic hot water, the heating system uses water as medium to heat indoor by ow temperature floor radiation, when we use the hot water in our daily life, the tap water heated by the heat exchanger in water tank and then we can use it by adjusting water mixing value.

B. Energy Recovery Ventilation

Through the provision of pipeline lines, the ERV can inhales the fresh outdoor air and discharges the gas that has been used in the room. And can be stored in the exhaust gas containing heat to preheat the outdoor air intake, so that it can reduce the number of residential energy consumption.

C. Photovoltaic System

Photovoltaic power generation is based on the principle of photovoltaic effect, the use of solar cells can be directly converted into solar energy. Produce electricity for daily use of electrical appliances or connected to the national grid. Photovoltaic panels mainly in roofing and wall.

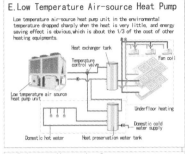

D. Ground Source Heat Pump

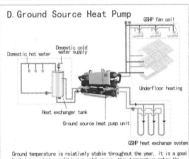

Ground temperature is relatively stable throughout the year, it is a good heat pump and air conditioning cold source, this temperature makes the ground source heat pump than the traditional air conditioning system operating efficiency is 40 %. So it save the energy and save about 40% of operating costs.

E. Low Temperature Air-source Heat Pump

Low temperature air-source heat pump unit in the environmental temperature dropped sharply when the heat is very little, and energy saving effect is obvious, which is about the 1/3 of the cost of other heating equipments.

F. Rainwater Harvesting System

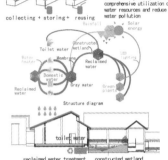

Rainwater recycling is an effective way to improve the comprehensive utilization of water resources and reduce water pollution.

This design combines rainwater recovery system with the artificial wetland, the collected rainwater was purified for toilet water, irrigation water and constructed wetland water.

Ecological Strategy

Passive Solar Energy Utilization
1. Solar house
2. Thermal storage wall
3. Heat storage floor
4. Directly benificial window
5. Movable shade panels
6. Warm porch
7. Natural ventilation system
8. Solar phase changed louver

Active Solar Energy Utilization
1. Solar water heating system
2. Photovoltaic system
3. Low temperature radiant floor

Other Technologies
1. Ground source heat pump
2. Low temperature aie source heat pump
3. Energy recovery ventilation
4. Rainwater harvesting system

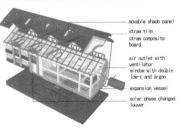
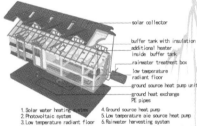

1. Solar water heating system
2. Photovoltaic system
3. Low temperature radiant floor
4. Ground source heat pump
5. Low temperature aie source heat pump
6. Rainwater harvesting system

Rendering

LEISURELY LIVING

综合奖·优秀奖
General Prize Awarded · Honorable Mention Prize

注 册 号：4947
项目名称：南山·颐居（西安）
　　　　　Zhongnan Mountain · Residence（Xi'an）
作　　者：黄浩、石涛、李光旭
参赛单位：吉林建筑大学
指导老师：肖景方

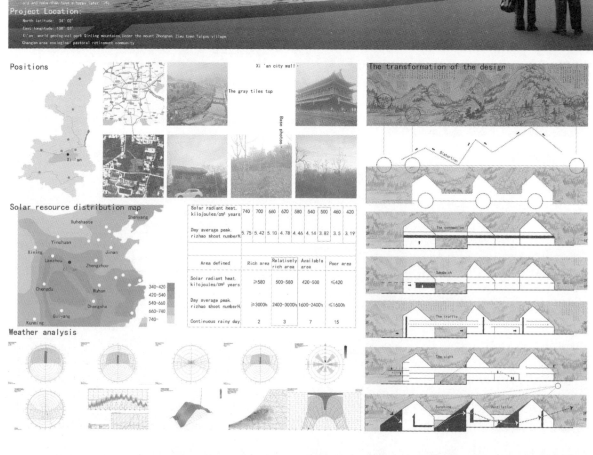

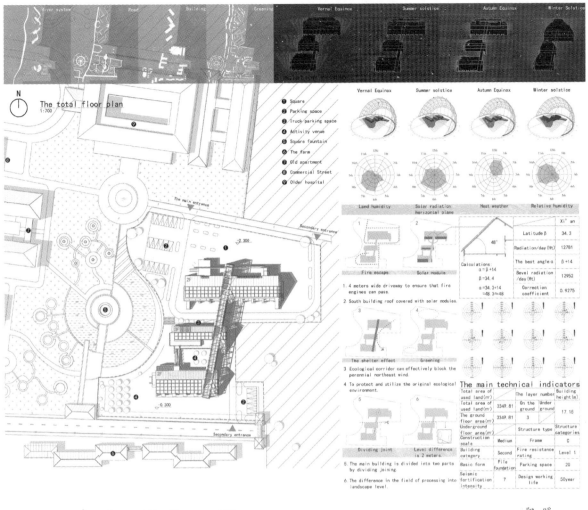

南山·颐居

【中国·西安】 ②

暖心温度人惬意，
阳灸全身叹自然。

——中国式养老

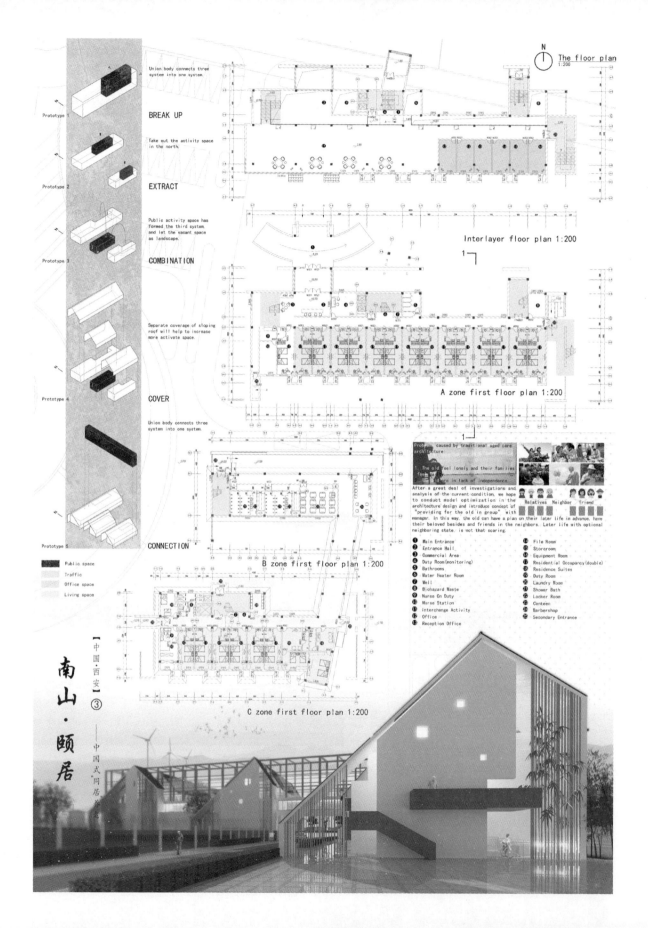

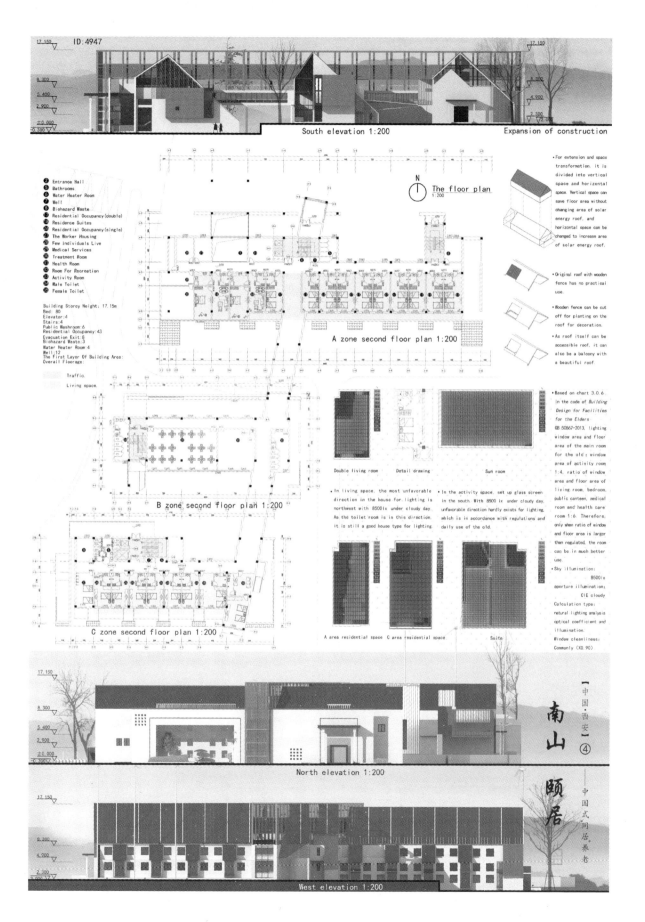

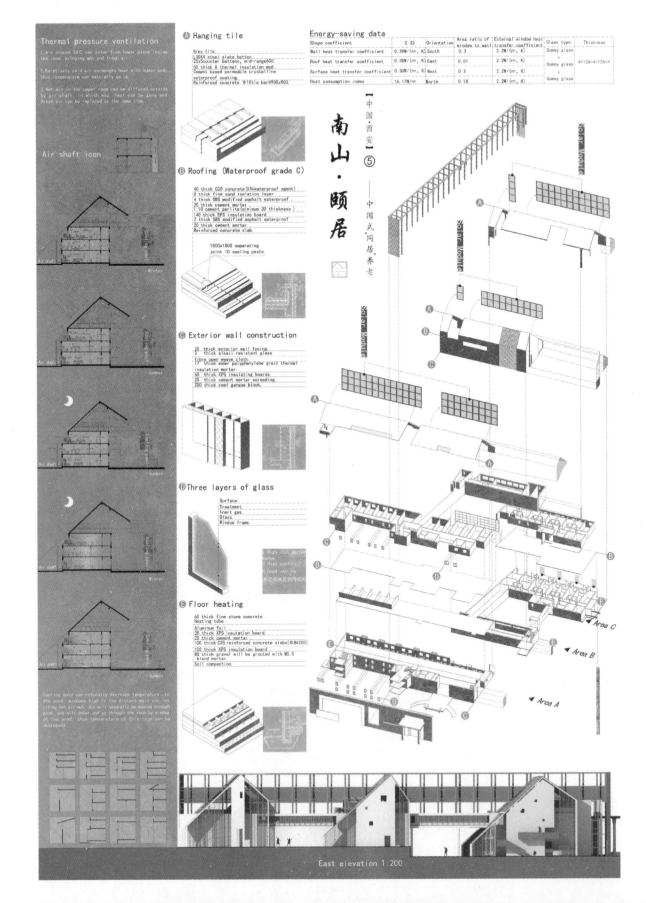

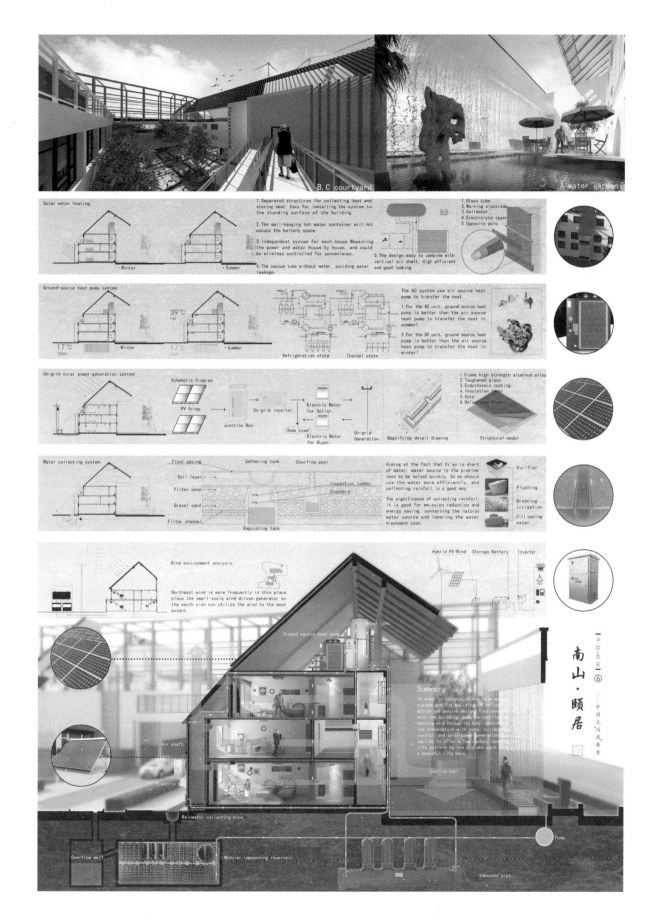

综合奖·优秀奖
General Prize Awarded · Honorable Mention Prize

注 册 号：5186
项目名称：秦岭·光居（西安）
Sunshine Elderly Care Center (Xi'an)
作　　者：张琮、于璐、陈萨如拉、陈孟栋
参赛单位：天津大学
指导老师：朱丽、汪丽君

秦岭·光居 01
Bird-view plan

Location Analysis
Xi'an is a long history city with the world famous historical and cultural, which is one of the four ancient civilizations in the world known to the world. 13 dynasties such as the Qin, Han and Tang set up the capital in here. The famous Silk Road starts from Xi'an. Terracotta Army, one of the eight wonders in the world, is located in Xi'an. The long history and cultural heritage makes Xi'an enjoy the reputation of "natural history museum".

Surrounding Analysis
Site lies on the Guanzhong Plain in central China, on a flood plain created by the eight surrounding rivers and streams. The city has an average elevation of 400 meters above sea level and an annual precipitation of 550 millimeters (22 in). The city borders the northern foot of the Qinling Mountains to the south, and the banks of the Wei River to the north.

Climate Analysis
Xi'an is warm temperate semi-humid continental monsoon climate, four distinct seasons, mild climate, moderate rainfall.

best orientation | annual incident solar radiation

weekly data (averge temperature) | weekly data (diffuse solar radiation)

hourly data | psychrometric

Natural and Culture

Design Description
设计从场地条件出发，将部分建筑嵌入地下，中和2m的高差，并平缓地将种植屋面与地面相连，通过体块的布置、下沉庭院、通高的阳光活动厅等形成丰富的空间层次和视线交流，为老年人提供交往的空间，减轻老年人的寂寞。在技术方面，我们综合使用了被动式技术和主动式技术。

Design from the site conditions, part of the building embedded in the ground, and 2 meters in height difference, and with a gentle planting roof will be connected with the ground. Through the arrangement of the body block, sink the courtyard and the high sunshine activity hall, the formation of rich spatial level and line of sight communication for the elderly to provide space for communication. In terms of technology, we have integrated the use of passive technology and active technology.

Behavior Analysis
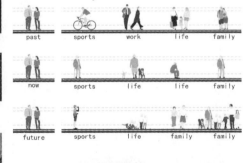

Advantage | Good medical facilities / A wealth of activities

Advantage | Care and companionship

Concept

site / at different level
connection / partly underground
sunlight room / add public space
daily life / qualitative spaces

Ground Floor Plan 1:300

2017 台达杯国际太阳能建筑设计竞赛获奖作品集

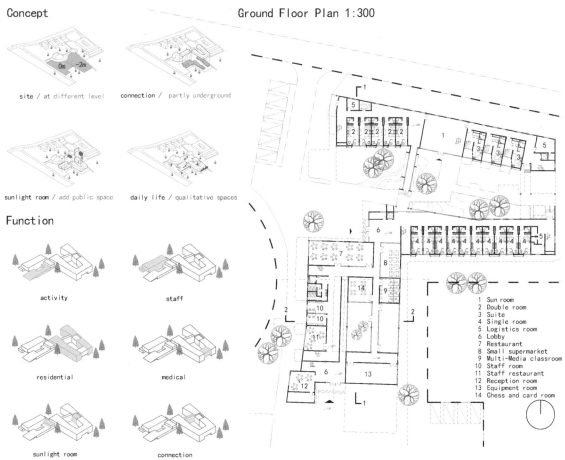

1 Sun room
2 Double room
3 Suite
4 Single room
5 Logistics room
6 Lobby
7 Restaurant
8 Small supermarket
9 Multi-Media classroom
10 Staff room
11 Staff restaurant
12 Reception room
13 Equipment room
14 Chess and card room

Function

activity
staff
residential
medical
sunlight room
connection

Shadow Analysis

winter solstice sun-path diagram   summer solstice sun-path diagram

Solar Environment Analysis

sunshine hours in winter   sunshine hours in summer

Wind Environment Analysis

comparison beween different shapes

Sunshine Elderly Care Center

秦岭·光居 02

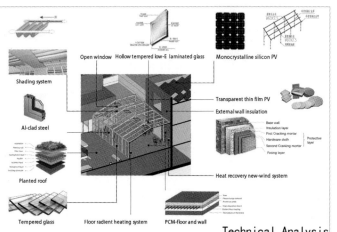

Technical Analysis
Comprehensive Application of Passive Method

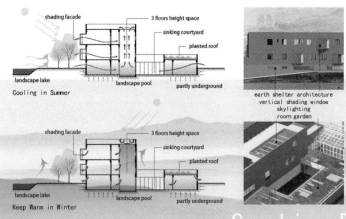

earth shelter architecture
vertical shading window
skylighting
room garden

Regulation of Sun Room Environment

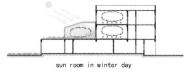
sun room in winter day

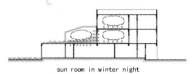
sun room in winter night

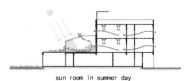
sun room in summer day

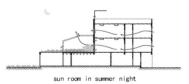
sun room in summer night

Second Floor Plan 1:400

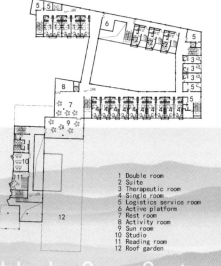

1. Double room
2. Suite
3. Therapeutic room
4. Single room
5. Logistics service room
6. Active platform
7. Rest room
8. Activity room
9. Sun room
10. Studio
11. Reading room
12. Roof garden

Sunshine Elderly Care Center

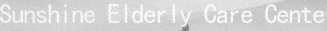

1-1 Section 1:200

秦岭·光居 03

Third Floor Plan 1:300

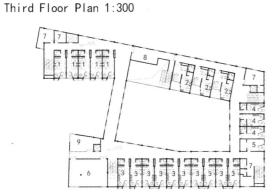

1 Double room
2 Suite
3 Single room
4 Therapeutic room
5 Laundry room
6 Multifunction room
7 Logistics auxiliary room
8 Active platform
9 Activity room
10 Office
11 Conference room

Sunshine Elderly Care Center

2-2 Section 1:200

Water Recycle System

Water recovery refers to the treatment of sewage, so that it can meet the appropriate water quality standards in a certain way. This technology is conducive to water conservation, improve the ecological environment.

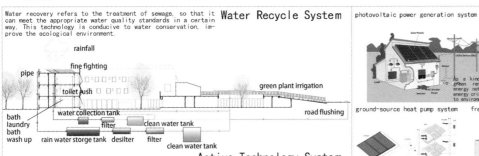

Active Technology System

Ground source heat pump can use underground shallow layer soil energy to keepcool in summer and warm in winter by consuming little high stand electrical energy to realize low stand energy transfer.
Solar hot-water system is a device which uses solar energy collector to collect solar radiation energy to heat water.

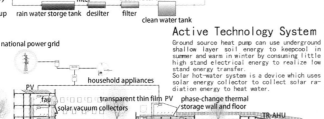

photovoltaic power generation system

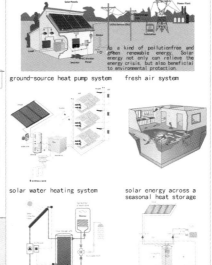

As a kind of pollutionfree and green renewable energy, Solar energy not only can relieve the energy crisis, but also beneficial to environmental protection.

ground-source heat pump system

fresh air system

solar water heating system

solar energy across a seasonal heat storage

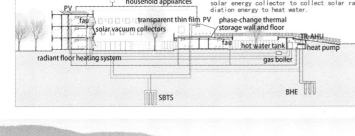

秦岭·光居 04

South Facade 1:200

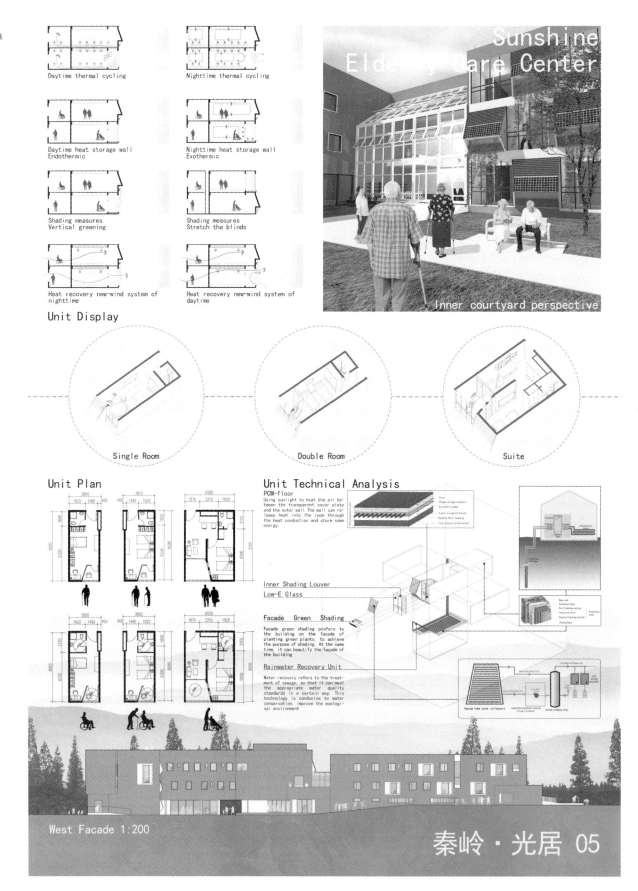

Sunshine Elderly Care Center

秦岭·光居 06

Site Plan 1:1000

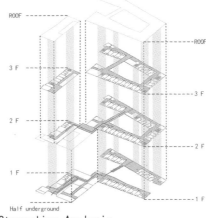

Streamline Analysis

Technical Analysis

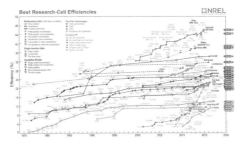

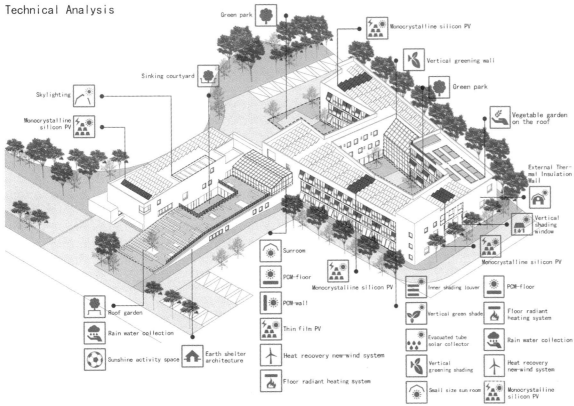

沐光·团居
Sunlight·Cluster

综合奖·优秀奖
General Prize Awarded · Honorable Mention Prize

注 册 号：5283
项目名称：沐光·团居（西安）
　　　　　Sunlight·Cluster（Xi'an）
作　　者：王　楠、王劲柳
参赛单位：天津大学
指导老师：刘丛红、杨鸿玮

District Analysis

Xi'an City Chang'an District, Zhongnan mountain, Taigou Village, located in Ecological Pastoral Community

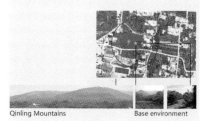

Qinling Mountains　　Base environment

Site Analysis

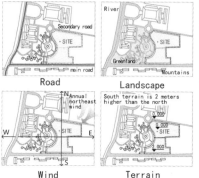

Road　　Landscape
Wind　　Terrain

Description

方案通过采用团居式的养老模式，将养护部分四个组团单元，老人通过组团居住的形式互帮互助，共享生活。单元采用预制装配式，尺寸使用模数。构件可以预制加工，考虑到后期加建。融入的阳光房、阳光活动厅、日光温室等，使老人沐浴在阳光里。坡屋顶、阳台集成了太阳能技术，光伏发电能够满足日常用电需求。场地内规划农园，采取雨水收集等节能措施。"沐光·团居"为老人营造更为舒适颐养、社区化、可持续的生态家园。

The design uses cluster living model, divide into four groups of units, the elderly help and share life through the cluster living form. The units use prefabricatio and the size is modulus. Components can be prefabricated, taking into account the late construction. The solar room, sunshine activity hall, solar greenhouse make the elderly bathed in the sunlight. Slope roof, balcony integrated solar technology, photovoltaic power generation meet the daily demand. Planning farm in site, take rainwater collection and other energy-saving measures. "Sunlight·cluster" create a more comfortable maintenance, community, sustainable ecological home for the elderly.

Concept Formation

Concept Deepen

According to the local weather conditions, make sure the planing layout and orientation to meet the comfortable wind environment, sunlight distance and sunlight condition of the activity field.

Best orientation　　Planning layout　　Wind frequency　　Wind environment

Sun Path simulation　　Sunlight shadow　　Solar radiation　　Sunlight distance

沐光·团居
Sunlight · Cluster

Concept Deepen

Road & function
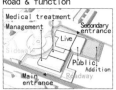
①Separation of pedestrian and vehicles. Reasonable function division. Consider addition part.

Landscape & view

②Block segmentation: beneficial to have good landscape view and form green gardens.

Sun & wind

③Adjust the orientation of the main rooms according to the best local orientation. And anti adverse wind.

Unit & group

④Residential unit module division: to divide four residential units forming group pension model.

Site plan 1:500

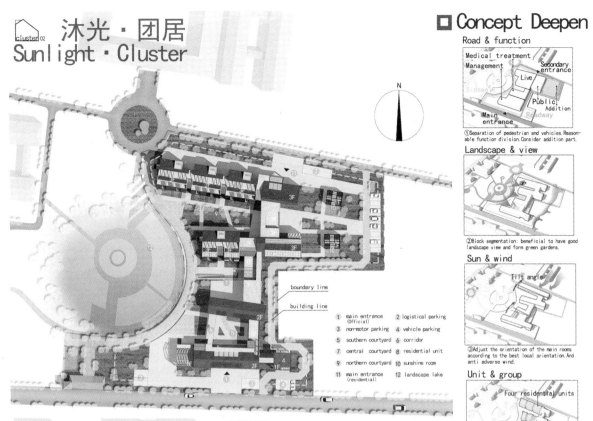

boundary line
building line

1 main entrance (Official)
2 logistical parking
3 non-motor parking
4 vehicle parking
5 southern courtyard
6 corridor
7 central courtyard
8 residential unit
9 northern courtyard
10 sunshine room
11 main entrance (residential)
12 landscape lake

First floor plan 1:200

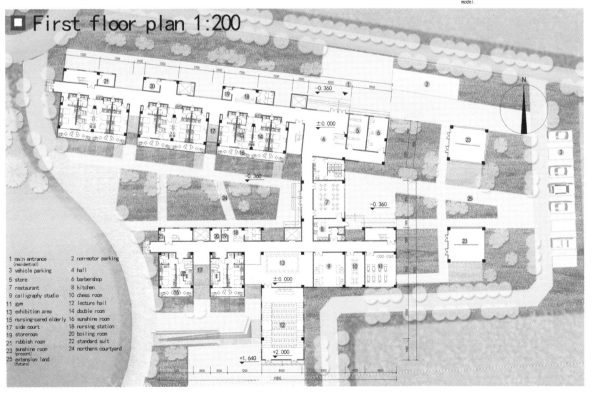

1 main entrance (residential)
2 non-motor parking
3 vehicle parking
4 hall
5 store
6 barbershop
7 restaurant
8 kitchen
9 calligraphy studio
10 chess room
11 gym
12 lecture hall
13 exhibition area
14 double room
15 nursing-cared elderly
16 sunshine room
17 side court
18 nursing station
19 storeroom
20 boiling room
21 rubbish room
22 standard suit
23 sunshine room (present)
24 northern courtyard
25 extension land (future)

沐光·团居
Sunlight · Cluster

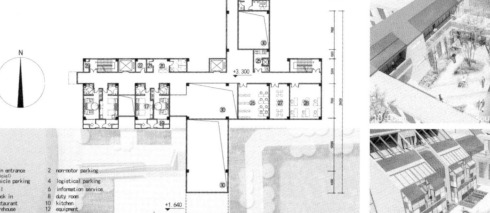

Legend:
1. main entrance (official)
2. nonmotor parking
3. vehicle parking
4. logistical parking
5. hall
6. information service
7. check in
8. duty room
9. restaurant
10. kitchen
11. warehouse
12. equipment
13. corridor
14. psychological room
15. healthcare room
16. southern courtyard
17. double room
18. nursing-cared elderly
19. sunshine room
20. nursing station
21. storeroom
22. boiling room
23. rubbish room
24. shower
25. servery
26. reading room
27. chess room
28. network room
29. sunshine room
30. over the space

▫ Second floor plan 1:500

▫ Elevation

South elevation 1:200

West elevation 1:200

Perspective details

9:00am — Lake trail

10:00 — southern courtyard

15:00pm — central courtyard

16:00pm — side court

17:00pm — southen atrium

沐光·团居
Sunlight · Cluster

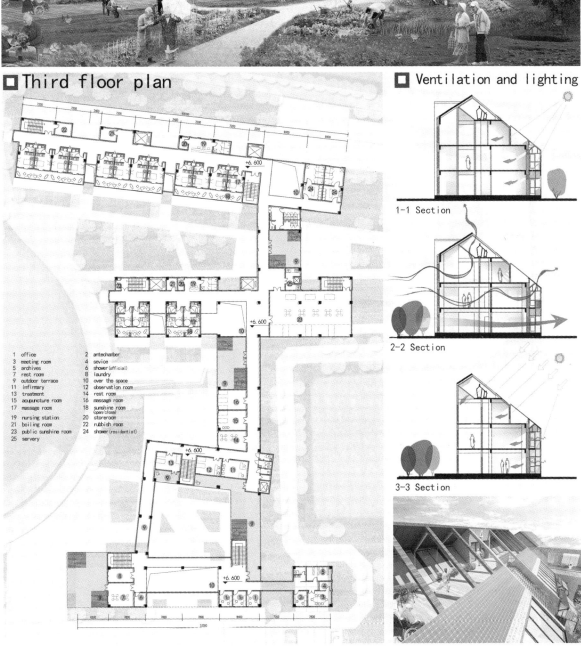

Third floor plan

1. office
2. antechamber
3. meeting room
4. sevice
5. archives
6. shower (official)
7. rest room
8. laundry
9. outdoor terrace
10. over the space
11. infirmary
12. observation room
13. treatment
14. rest room
15. acupuncture room
16. massage room
17. massage room
18. sunshine room (open/close)
19. nursing station
20. storeroom
21. boiling room
22. rubbish room
23. public sunshine room
24. shower (residential)
25. servery

Ventilation and lighting

1-1 Section

2-2 Section

3-3 Section

沐光・团居
Sunlight・Cluster

◻ Living cells

Different types of the old people have different living status and habits. A variety of houselayouts so as to adapt to different needs.

Layout A 👥+👥
Inhabitants: 2 disability elderly + 2 healthy elderly
Area: 52m² number: 6
Orientation: South

Layout B 👥+👥
Inhabitants: 2 disability elderly + 2 healthy elderly
Area: 52m² number: 2
Orientation: South

Healthy elderly canlook after the disability elderly especially in the accident

Layout C 👥+👥
Inhabitants: 1 couple elderly + 2 elderly
Area: 52m² number: 6
Orientation: South

Couple elderly can live together like at home. They can cook or work and help other elderly.

Layout D 👥
Inhabitants: suite 2 elderly
Area: 56m² number: 2
Orientation: South

Prosperous old people can live in comfortable suite, with viewing balcony

Layout E 👤+👤
Inhabitants: single room 1+1 elderly
Area: 32m² number: 8
Orientation: South

The old people living alone can still go to activity platform to communicate with others

Humanized Design

- Movable partition wall
- Accessible toilet
- Semi-private entrance
- Flexible kitchen and office
- into community area
- Integrated storage
- Sharing balcony

◻ Prefabricated units

Design project use "housing group" old-age security model, dividing into four groups of units: NO.1-4. The units adopt prefabrication with modulus size. Column, beam, floor, staircase and balcony are prefabricated, they will be built on location. Exterior, roof and balcony integrate solar energy utilization technology.

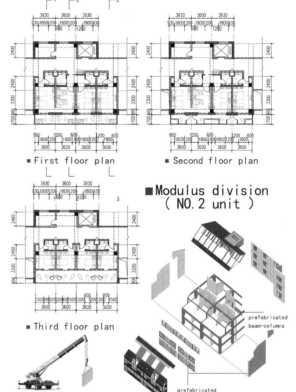

- First floor plan
- Second floor plan
- Third floor plan

■ Modulus division (NO.2 unit)

- prefabricated beam-columns
- prefabricated insulation walls
- transportation and build on-site

Prefabrication System

◻ Solar energy use

- solar room
- temporary greenhouse
- sun activity hall

■ Prefabricated solar room

The solar room is hollow through two floors, it willmake full use of the sun and its main effect is to transfer heat to bedrooms. The old people can bask themselves on the sunny balciny. The old people can basked ourselves on the sunny beach

Roof angle is chosen as 38°, solar panels have the most efficient to use the solar energy.

- Opening windows for ventilation
- Integrated solar photovoltaic panels
- Sun-shield for window shade
- Planting groove for balcony planting
- Structure of technical support
- Prefabricated insulate wall to storage heat

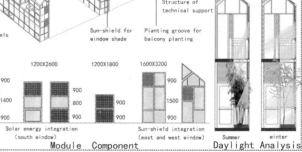

- 1200X3200
- 1200X2600
- 1200X1800
- 1600X3200

Solar energy integration (south window) | Sun-shield integration (east and west window)

Module Component | **Daylight Analysis** (Summer / winter)

沐光·团居
Sunlight · Cluster

■ Sun Activity Hall

Activity hall for the old   Greening Variable sector
Sun activity hall can reach effect of regulating temperature humidity through the varible window, visor and indoor green planting.

■ Temporary greenhouse

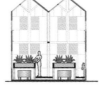

The late construction area
Now: temporary greenhouse
Future: temporary greenhouse dismantle, component will be used to construct residential unit solar room.
Temporary greenhouse can be dismantled. The componet used for future construction
Greenhouse planting activities rich old people's life

■ Solar water-heating system

BIPV panels
Living Space   Management and Office
①BIPV panels ②controller ③water supply ④thermal storage tank
⑤cold water input ⑥hot water output ⑦hot shower ⑧domestic hot water

■ Solar Chimney

Appearance   Solar chimney ventilation diagram

Solar chimney used hot and wind pressure difference for ventilation. Form and color of the tower develop by the xi'an tower.
simple air flow   solar chimney

□ Insulated wall

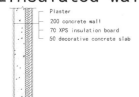

Plaster
200 concrete wall
70 XPS insulation board
50 decorative concrete slab

Prefabricated insulated wall have the effect of heat preservation and heat insulation, reduce building energy consumption.
Concrete column
Wall reserved reinforcement
Waterproof sealing material
Connecting piece

Prefabricate connection

□ Geothermal heating

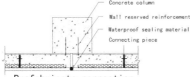

Fan coil unit   main engine

Absorption of soil temperature

The local has rich geothermal resources, the project make full use of geothermal, set the ground source heat pump for floor heating.

Geothermal heating

■ Solar radiation & collection

Rooftop solar panels

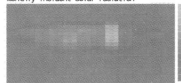

Annual solar radiation distribution

Monthly incident solar radiation

Solar panels' hourly collection

Comparison of using and collection

Simulations statistics show that the 658 m²solar collector area's annual cumulative solar generation is about 39000 kWh. According to the data of China electric power federation, domestic consumption of per person in Shaanxi province is about 395kwh, the cumulative used electricity is about 37500kWh (using number 80 beds and service workers). Solar power generation is greater than the usage throughout the year.

□ Rain water collection

road runoff   roof planting   roof planting   balance tank equipment   wash the car
rain collection   rain collection   rain conversion process
flush toilet   agricultural irrigation   flush toilet   waterscape
middle-water use

Rainwater collecting recycling system (RCR system)

Rain vertical pipe
Rainwater collection device
Balcony Planting
Rain vertical pipe

Crop
Primary soil
Sand
Geomembrane package the rain
Inlet pipe
Rainwater collecting box
Outlet pipe
Sand
Primary soil

Balcony rainwater collection diagram   Rain infiltration facilities

温暖的房子 01
Warm House

综合奖·优秀奖
General Prize Awarded · Honorable Mention Prize

注　册　号：5440
项目名称：温暖的房子（西安）
　　　　　 Warm House（Xi'an）
作　　者：张　玺、倪晨辉、孙宁晗、
　　　　　 张子奇、万华楠、毛晓天、
　　　　　 韩赟聪
参赛单位：山东建筑大学、北京建筑大学
指导老师：赵学义、崔艳秋、欧阳文

设计说明

该方案充分考虑了西安当地的气候特点与老年人平日生活的特点，方案中采用了街巷、连续坡屋顶、庭院等传统西安民居元素。并结合老年人的心理特征设计了阳光间、种植区、温泉浴场、活动空间等供老年人休闲使用的空间。方案采用绿色建筑设计，运用了多种绿色建筑技术如地源热泵、屋顶墙面绿化、太阳能烟囱、雨水收集系统、中水处理系统、双层幕墙系统等绿色建筑技术。

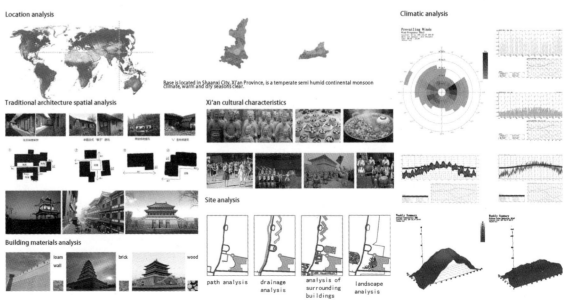

温暖的房子 02
Warm House

PLANE THERMAL RADIATION ANALYSIS

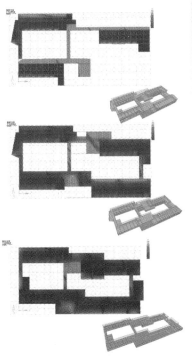

Design Description

The scheme fully consider the characteristics of Xi'an local climate characteristics and daily life of the elderly, the method of streets, continuous sloping roof, the courtyard of Xi'an traditional folk house element. Combined with the psychological characteristics of the elderly design of the sun, planting area, spa, activity space for the elderly leisure space. The scheme of green building design, the use of a variety of green building techniques such as ground source heat pump, solar chimney, wall greening, roof rainwater collection system, water treatment system, double curtain wall system of green building technology.

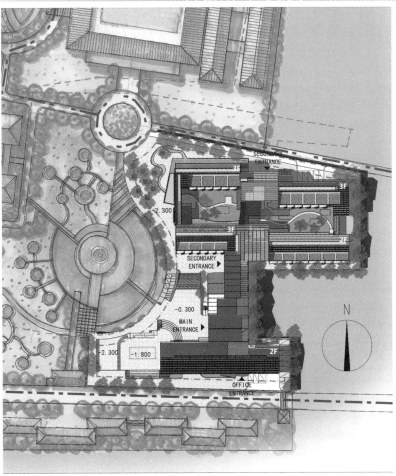

SITE PLAN 1:500

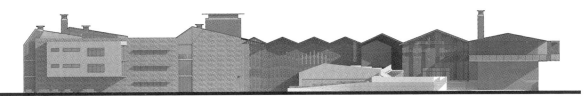

WEST ELEVATION 1:200

SOUTH ELEVATION 1:200

温暖的房子 03
Warm House

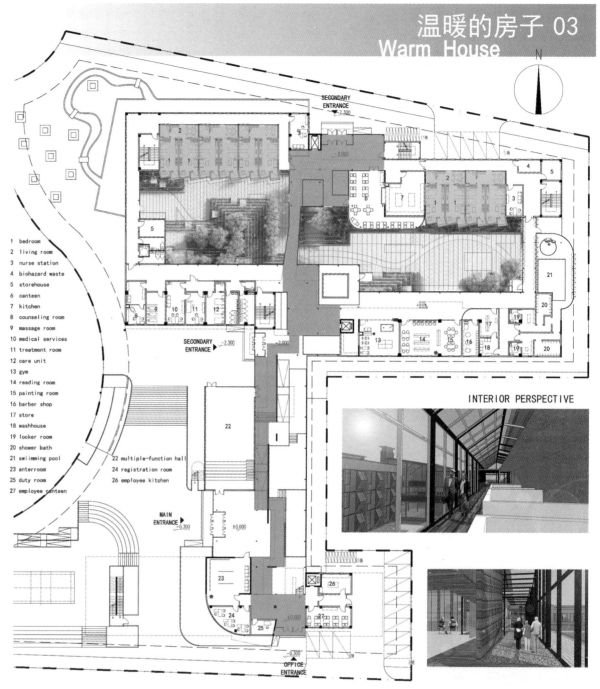

1 bedroom
2 living room
3 nurse station
4 biohazard waste
5 storehouse
6 canteen
7 kitchen
8 counseling room
9 massage room
10 medical services
11 treatment room
12 care unit
13 gym
14 reading room
15 painting room
16 barber shop
17 store
18 washhouse
19 locker room
20 shower bath
21 swimming pool
22 multiple-function hall
23 anteroom
24 registration room
25 duty room
26 employee kitchen
27 employee canteen

INTERIOR PERSPECTIVE

1ST FLOOR PLAN 1:200

SECTION 1:200

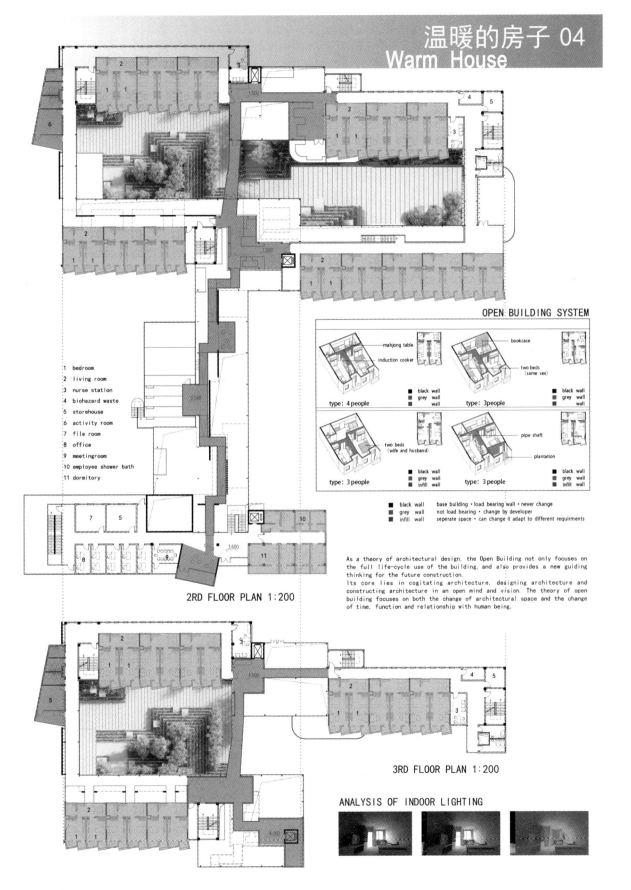

温暖的房子 05
Warm House

COURTYARD

THE SUMMER SOLSTICE SUN SHADOW MAP ANALYSIS

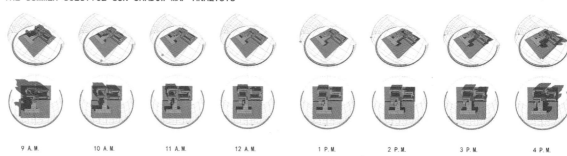

9 A.M.　　10 A.M.　　11 A.M.　　12 A.M.　　1 P.M.　　2 P.M.　　3 P.M.　　4 P.M.

Horticultural Therapy

HT(Horticultural therapy), which had originated in the end of 17th century, is an intercross subject that comprises horticulture, medicine and psychology. It was proved to have different effects on patients, such as the disabled, mental disease, low IQ and the aged. It is a supplement and improvement to the modern medicine.

spatial analysis

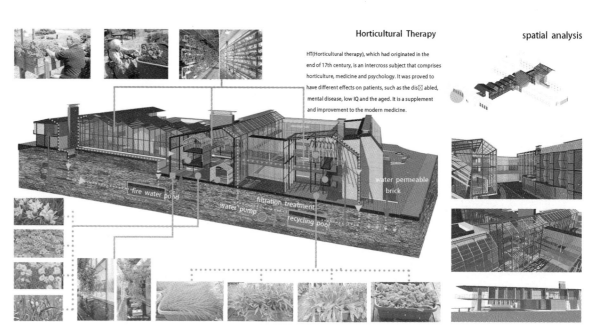

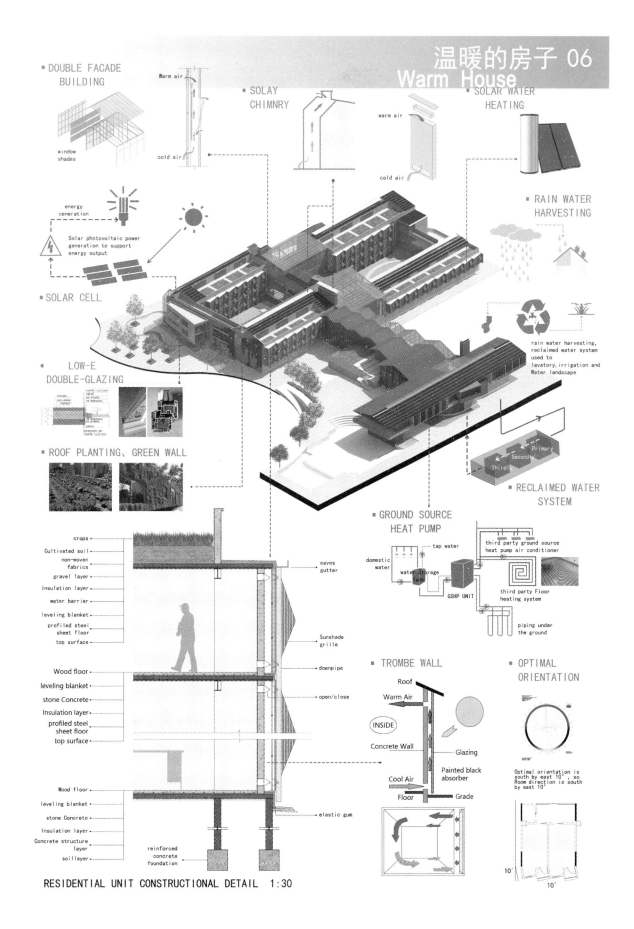

园居安老
西安生态颐养服务中心
Pastoral pension center

综合奖·优秀奖
General Prize Awarded · Honorable Mention Prize

注 册 号：5458
项目名称：园居安老（西安）
　　　　　Pastoral Pension Center (Xi'an)
作　　者：徐笑、白梅、潘莹莹、钟江龙
参赛单位：华中科技大学
指导老师：无

Site Location Analysis

Shang Xi　　Xi'an

This project is located at Chang'an District, Xi'an City, Shaanxi Province. The land of the project has thick vegetation and the landscape is beautiful. It is suitable for the elderly to live here and enjoy their life.

Changes of Life After Retirement

	Young People	The Elderly	
	Working Person	Retiree	
Main activity areas	Work unit	Family	Main activity areas
Social Communication	Colleagues	Family and neighbors	Social Communication
Pace of Life	According to their work	Self-determination	Pace of Life
Knowledge Skills	Work content	Their own interests	Knowledge Skills
Roll Play	Social role	Self role	Roll Play

Xi'an Weather Data

1. The optimal orientation of building is south
2. The temperature, wind speed, precipitation, sunshine situation summary map
3. The Amount of Solar Radiation

4. Average Temperature
5. Yearly Humidity Change
6. The Comparative Analysis of the Four Seasons

Physiological Characteristics of The Elderly

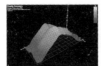

1. It's difficult to distinguish similar color for the elderly
2. The elderly move with difficulty.
3. The old people are increasingly likely to fall.

4. Temperature changes always bring the elderly with chronic diseases.
5. The elderly need to bask in the sun.
6. It's difficult for the old man to reach high place.

设计说明：

考虑到基地所处的区位风貌，采取传统与现代相结合的建筑风格。以传统四合院为原型，运用庭院式布局，将自然与建筑结合，为老年人创造丰富的室外活动空间。在生态方面，被动式与主动式相结合。被动式设计方面采用建筑朝南，设置庭院和阳光房，对侧开窗、加强围护结构保温等的方式。主动式技术方面采用太阳能系统、雨水收集系统和地热系统。

Design Notes:

In consideration of the characteristics of the region, we combine traditional architectural style with modern ones to design the building. By taking the traditional Siheyuan as a prototype and using the design method of integral layout with courtyard type, the aged center forms more outdoor activities space. In terms of ecology, passive and active techniques are combined. In passive design, the building is designed to face south. The courtyard and the sun room are set up for heat storage. Enhance the heat-insulating property of building envelope by some methods. Solar energy system, rainwater collection system and undersoil heating system are adopted in active design.

园居安老
西安生态颐养服务中心

Pastoral pension center 2

Conception Digram

1. Put a L-shaped block according to the red line.

2. Cutting the block according to the altitude difference.

3. Cut the blocks into long strips to get as much sunlight as possible.

4. Create the entrance space by push-and-pull of the blocks.

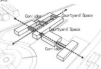
5. Use the corridors to connect all the blocks and we can get two courtyards for the elder to do some outdoor activities.

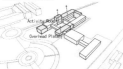
6. Increase appropriately number of floors with consideration of the building area and internal functions. And part of the building is built on stilts to create grey space.

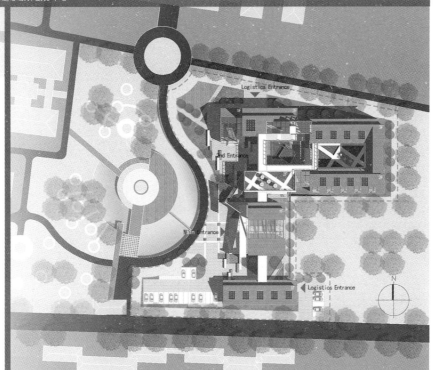

Site Plan 1:500

Energy Saving of Building Envelope

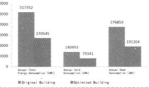

Sweal is used for simulating a rough energy-saving calculation. The results show that the energy saving ratio of the total energy consumption of the building is 47.7%. Meanwhile the ratio of the annual consumption is 43.53% and the ratio of the annual heat consumption is 49.26%. The optimized building is better than the original building with a external wall insulation structure, roof insulation structure, double glass curtain wall and double hollow glass windows.

Econo-technical Norms:

Site Area: 10282 ㎡
Green Ratio: 41%
Total Floor Area: 3338 ㎡
Floor Area Ratio: 0.32
Building Density: 21%
Building Coverage Area: 2196 ㎡

The Shadow Range

Summer Solstice 9:00 | Summer Solstice 11:00 | Summer Solstice 13:00 | Summer Solstice 15:00

Winter Solstice 9:00 | Winter Solstice 11:00 | Winter Solstice 13:00 | Winter Solstice 15:00

Summer Shadow Range — The solar elevation angle is high in the summer solstice, so the building has less architectural shelter. Shadow in 9:00-15:00 does not changes obviously. It has a litter influnce on the building.

Winter Shadow Range — The solar elevation angle is low in the winter solstice, so the building has more occlusion. Shadow in 9:00 -15:00 changes significantly. It has a great influnce on the building.

West Elevation 1:200

园居安老
西安生态颐养服务中心

Pastoral pension center 3

Bird's Eye View Drawing

Ground Floor Plan 1:300

Functional Zoning Map

- Living room
- Management service space
- Medical room
- Activity room
- Traffic space
- Green space
- Sunshine hall
- Living support room
- Vertical traffic

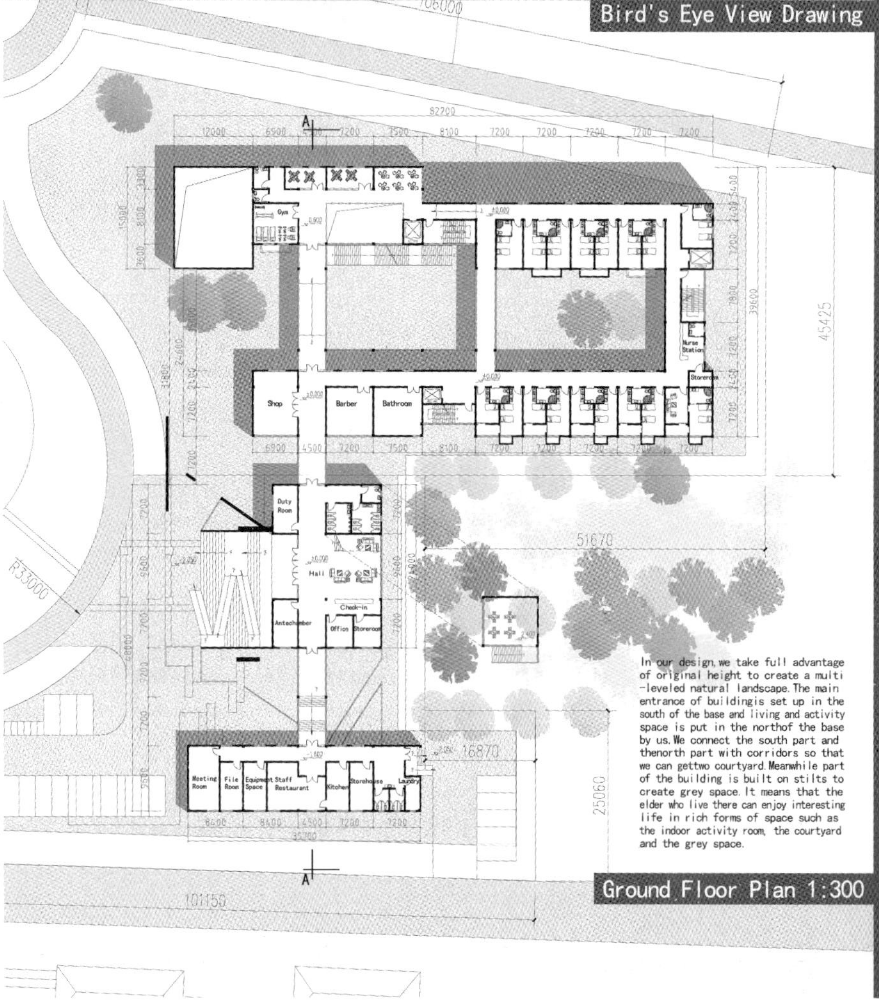

Streamline Chart

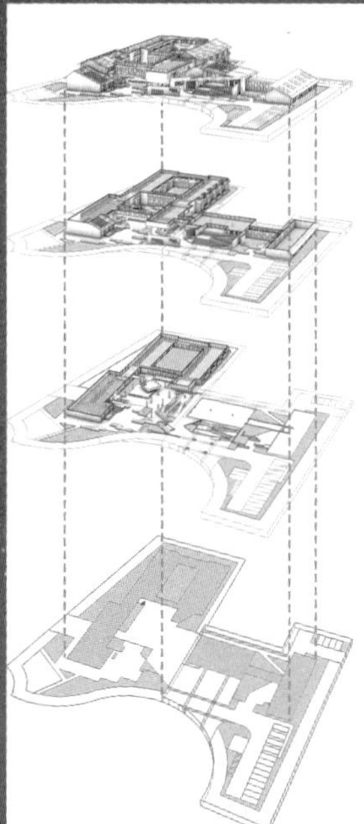

In our design, we take full advantage of original height to create a multi-leveled natural landscape. The main entrance of building is set up in the south of the base and living and activity space is put in the north of the base by us. We connect the south part and the north part with corridors so that we can get two courtyard. Meanwhile part of the building is built on stilts to create grey space. It means that the elder who live there can enjoy interesting life in rich forms of space such as the indoor activity room, the courtyard and the grey space.

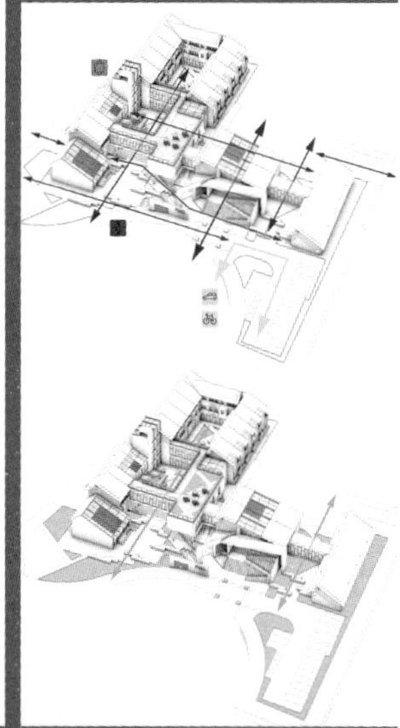

With the limitation of elders' mobility, they rely much on the facilities on the ground and they love to live with nature in a more idyllic way. So we design the building with low density and high green rate.

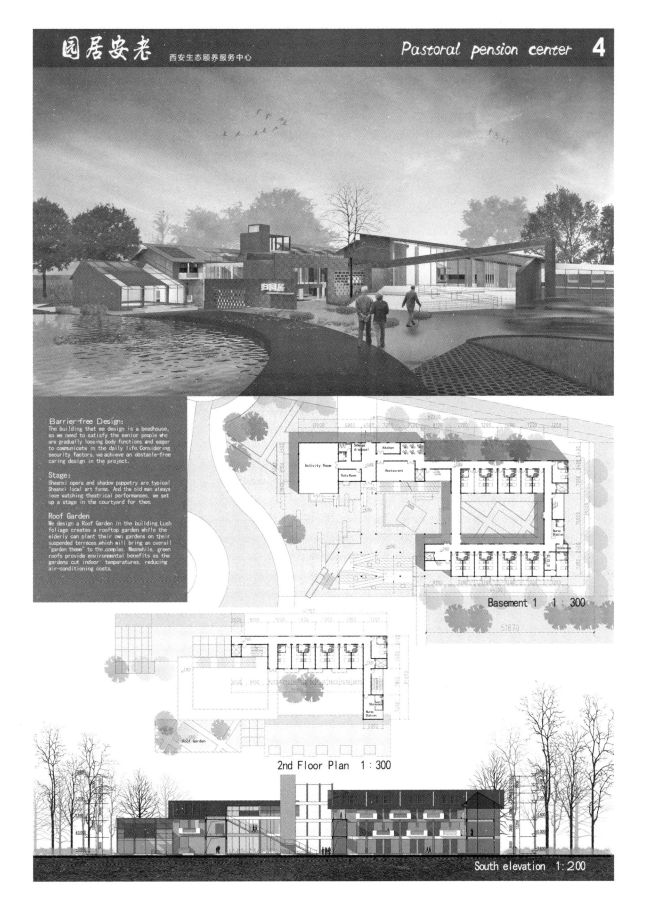

园居安老 西安生态颐养服务中心 — Pastoral pension center

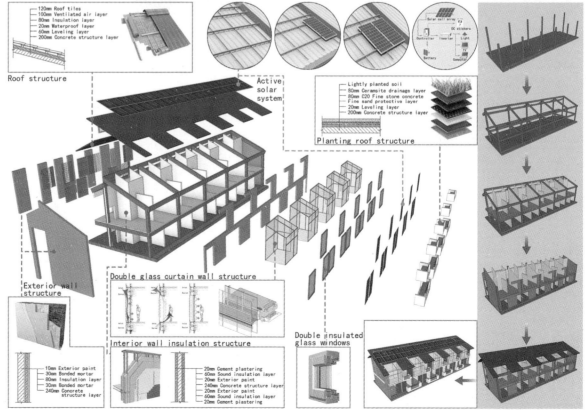

Building Explosion Map

Construction Process

园居安老 西安生态颐养服务中心

Pastoral pension center

Lighting Factor Analysis

Radiance is used for simulating lighting factor of the interior. The main space lighting factor of the old-age building needs to reach 3% to meet the needs of the elderly. The simulation results show that the lighting factor of most areas reaches 3%, 13% of the first layer, 12% of the second floor, only 4% of the three buildings do not meet the requirements of the building lighting factor. The areas are the washroom, walkway, elevator which have low lighting requirements. Therefore, the daylight of building basically meet the lighting requirements.

Perspective

Sunshine Time

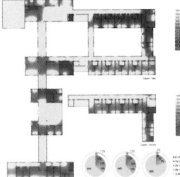

Ecotect is used for simulating the sunshine time. According to the specification, the old-age building should meet the requirements that sunshine time of the Living space is not less than 2h on the winter solstice. The simulation results show that the building meets the sunshine requirements.

Natural Ventilation

Opening the opposite side of the window is conducive to air circulation. Side of the high windows and the opposite side of the low window is conducive to the flow of the room the wind, creating a comfortable wind environment.

Summer Natural lighting   **Winter Natural lighting**   **Day Natural lighting**   **Night Natural lighting**

By using the difference of the solar altitude angle in summer and winter, summer shade, while winter heat.

When the sun shines into the room during the day in winter, the external walls and floor store heat. And release it gradually during the night.

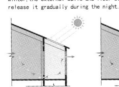

Winter Day   **Winter Night**   **Summer Day**   **Summer Night**

Open the inlet and the outlet during the day in winter. Then the heated air in the sun room get into the room. Close the inlet and the outlet when it is evening. Then the heated air stay indoors.

Close the upper outlet when it is summer. Strengthen the indoor ventilation, and improve human thermal comfortable.

A-A section 1:200

综合奖 · 优秀奖
General Prize Awarded · Honorable Mention Prize

注 册 号：4549
项目名称：逸院昀寮（泉州）
The Wander over Sunshine (Quanzhou)
作　者：徐明哲、耿煜周、吴　寰、赵　亮
参赛单位：天津大学、湖南大学、深圳大学
指导老师：无

THE WANDER OVER SUNSHINE
逸院昀寮

设计说明

方案选址于泉州市德化县雷峰镇瓷都印象生态园，基于当地气候，融合泉州传统民居手巾寮的建筑智慧，设计适老、健康、舒适、生态、环保的绿色空间。建筑整体采用合院式的布局，有利于老人之间交流活动。为了考虑老年人颐养以及公共服务设施与整个园区的共享，居住生活用房被安排在二层以上，同时融入护士站、阳光房、庭院、晒台、屋顶花园等多项服务空间。建筑内外全部采用无障碍设计，营造一个安全的康养环境。建筑整体融入太阳能设计，每个居住单元上都有兼具通风、隔热、遮阳、太阳能发电的复合屋顶，无论冬夏早晚，均可最大限度得利用阳光，创造阳光、颐养的生态居所。

DESIGN DESCRIPTION

The project is based on the local climate where is in Quanzhou Porcelain Impression Ecological Park, integrating traditional residential architectural wisdom-Jin Liao of Quanzho.It aims at designing appropriate old, healthy, comfortable, ecological, green space. The whole building adopts the layout of the courtyard, which is conducive to the leisure activities between the elderly. In order to consider the elderly and the maintenance of public service facilities and the entire park, the living space is arranged in the second floor, and they are surrounded with the nurses station, sun room, courtyard, balcony, roof garden and many other service space. The whole building uses barrier-free design to create a safe environment for health care and convenience. The project use the solar energy design as well, each living unit has ventilation, heat insulation, shade, solar solar power of the composite roof, both in winter and summer. All strategies can be used to maximize the sun, creating sunshine, maintenance of ecological residence.

CONCEPT

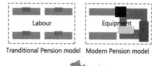

Tranditional Pension model　Modern Pension model　Ecological Pension model

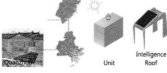

Quanzhou　Unit　Intelligence Roof　Ecological Room

CLIMATE ANALYSIS

Summary

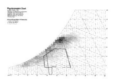

From the enthalpy humidity chart and the best orientation map, the passive solar energy and geothermal energy are more suitable for the passive technology. The best orientation should be 5 degrees south east.

Solar Radiation

From the hourly data, the direct solar radiation from seven to August and from December to October was higher, and the amount of solar diffuse radiatio was higher in the period from February to June.

Sunshine Direction

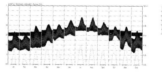

North　East　Southeast
South　West　Southwest

From the point of view of the amount of solar radiation in different directions of the year, west, southeast and southwest radiation is higher, and two seasons have greater fluctuations in the autumn and winter.

Wind Frequency

Spring　Summer　autumn　winter

According to the wind frequency and direction from the four seasons, spring and summer wind direction is east while autumn and winter wind direction is southwest.

Active strategy:
1 Active solar panels (solar collectors), for the construction of electricity (hot water, heating)
2 Green Roof and vertical plants.

Passive strategy:
1 Best orientation for the south;
2 Set the sun room next to the elderly living units;
3 Passive ventilation roof, reduce the roof to absorb the radiant heat, to avoid the top floor temperature rise;

Conclusions

Combined with local climate and building functional requirements, we can make full use of solar energy and ventilation conditions.

THE WANDER OVER SUNSHINE

MASTER PLAN 1:500
TECHNICAL AND ECONOMIC INDICATORS
Building Site Area: 4927.65m²
Building Area: 3373.27m²
Floor Area Ratio: 0.79
Building Density: 36.6%
Greening Rate: 24.5%
Ground Parking: 6
Barrier Free Parking: 1
Building Storey: 3
Building Height: 14.5m

SITE ANALYSIS

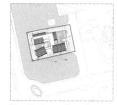

The East Entrance
The courtyard of the master plan is just in line with the park in the east of the main road.

The Western Boundary
The central courtyard is open to the west, being harmony with the vegetation.

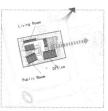

Traffic Organization
There are northern, southern, eastern three entrances for pedestrians, in order to keep in directly touch with the zone.

Function Layout
Except for the provides for the aged, the design also serves for the whole zone. The living rooms, medical rooms and public activity rooms are open to the zone, which are located at the transportation convenient place.

SITE COMPUTING

Insolation Levels

Total Radiation 120000-1200000 Wh   Average Daily Radiation 320-3220 Wh

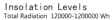

From the simulation of Ecotect, the building's annual solar radiation and daily solar radiation the basically meet the needs of the service center.
Northen corner of the building have less radiation where arranged the toilet, equipment and other service space,,
The atrium where have lower heat radiation is not surrounding with the main function.

Shadow Range

The Winter Solstice   The Summer Solstice

The winter sunshine are met through the adjustment of the layout of the building.

Because of the blocking of the active center, enough shade can be provide in summer.

NORTHEAST VIEW

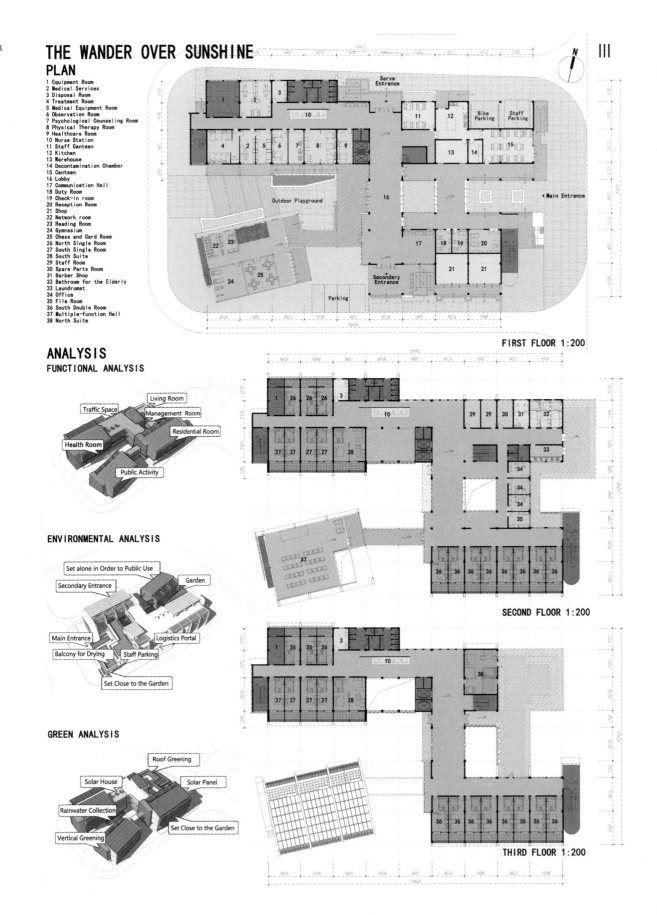

THE WANDER OVER SUNSHINE

IV 2017 台达杯国际太阳能建筑设计竞赛获奖作品集

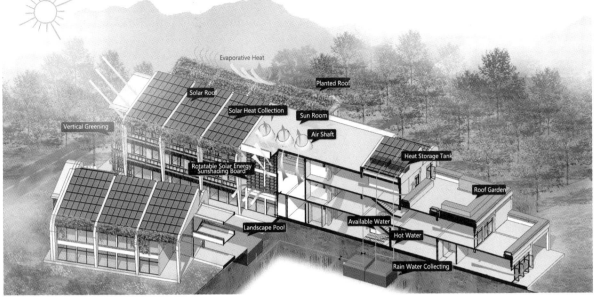

☀ SOLAR ENERGY PROJECT

System Profile (Whole Year)

Water Load (Kw.h)	57010.2
Solar Guaranteed Rate (%)	97.3
Load Energy Rate (%)	139.4
Dynamic recovery cycle	13.0 years

Photothermal Profile (Annual Total)

Collector Area (m²)	270.0
Solar radiation Amount of Collector Surface (Kw.h)	443855.0
Collector Heat Gain (Kw.h)	180075.5
Loss Energy of Pipeline and Tank (Kw.h)	22080.7
Auxiliary Heating Energy (Kw.h)	1552.5
Carbon Dioxide Emission Reduction in SystemLife	151.4 t

Heat Guaranteed Rate of Solar

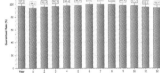

System Principle

System Name	Constant Temperature Water + Temperature Difference Circulation + Direct Heat Exchange + Non Pressure Water Tank + Direct Heating System
Heat Collection and Heating Range	Centralized Heat Collection, Centralized Heat Storage and Centralized Heating
System Operation Mode	Forced Circulation

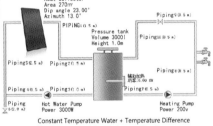

Constant Temperature Water + Temperature Difference Circulation + Direct Heat Exchange + Non Pressure

Economic Environmental Benefit

Economic Analysis

Solar Hot Water System Investment	279150.0 yuan
Saving Amount of Solar Hot Water System Annually	199647.6 MJ
Total Cost Savings During System Life	30061.0 yuan
Simple Payback Period	8.5 years
Dynamic Payback Period	13.0 years
Carbon Dioxide Emission Reduction in System Life	151.4 t

Contrast of Cost

🪟 INSOLATION OF SHUTTER

Rotatable Shutter | No Rotatable Shutter

320-3290 Wh Average Daily

450-4140 Wh Average Daily

210-1540 Wh Summer Average Daily | 450-4550 Wh Summer Average Daily

210-1730 Wh Winter Average Daily

210-1920 Wh Winter Average Daily

The simulation results shows that the installation of shutterss have great impact on greater heat radiation. It can play a flexible adjustment according to the needs of indoor heat radiation role.

EFFECTS OF ROOF GARDEN

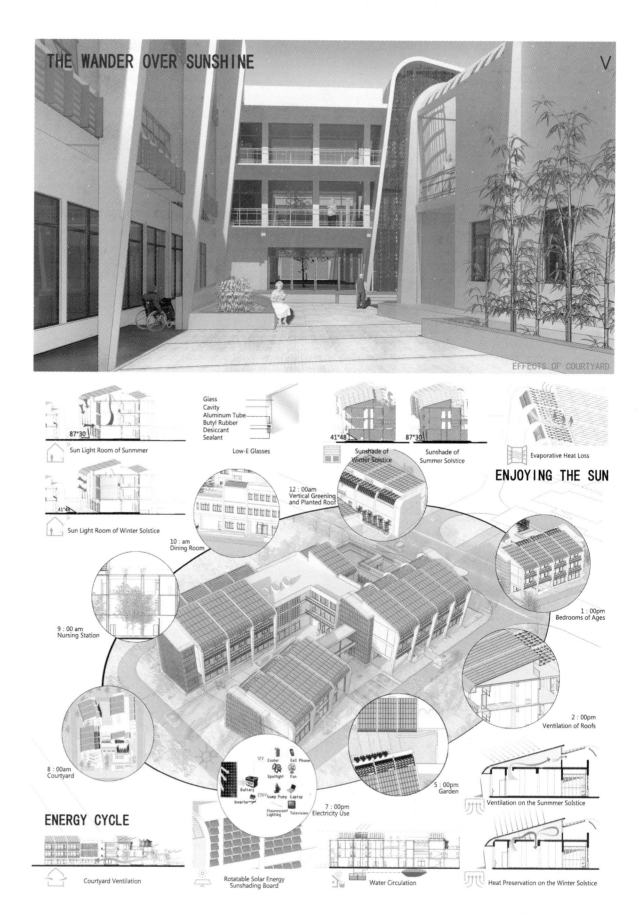

THE WANDER OVER SUNSHINE

SLAB SECTION

ELEVATION

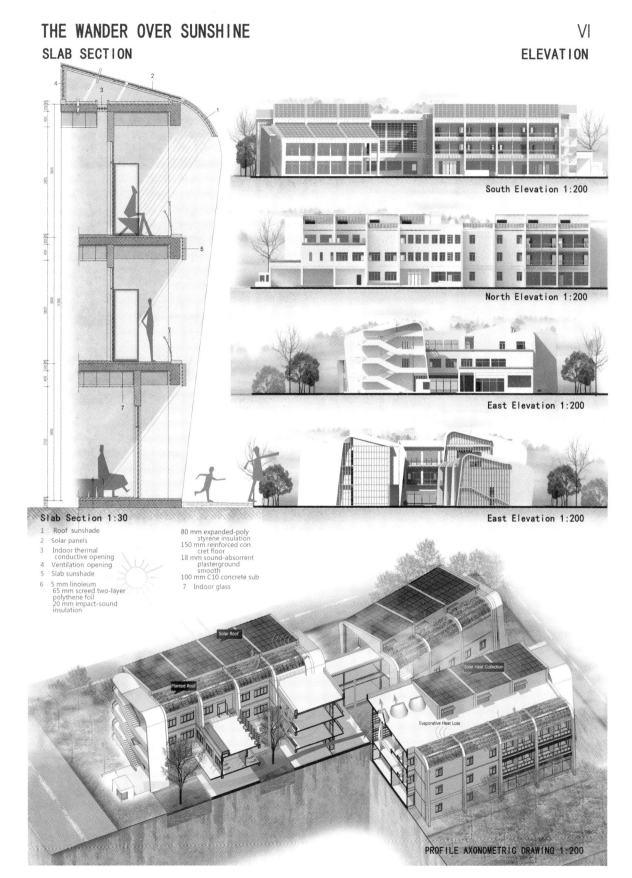

Slab Section 1:30
1. Roof sunshade
2. Solar panels
3. Indoor thermal conductive opening
4. Ventilation opening
5. Slab sunshade
6. 5 mm linoleum
   65 mm screed two-layer polythene foil
   20 mm impact-sound insulation
   80 mm expanded-polystyrene insulation
   150 mm reinforced concret floor
   18 mm sound-absorrent plasterground smooth
   100 mm C10 concrete sub
7. Indoor glass

South Elevation 1:200
North Elevation 1:200
East Elevation 1:200
East Elevation 1:200

PROFILE AXONOMETRIC DRAWING 1:200

综合奖·优秀奖
General Prize Awarded · Honorable Mention Prize

注 册 号：4770
项目名称："院"儿里院外（泉州）
　　　　　In Yard & Out Yard
　　　　　（Quanzhou）
作　　者：赵梓汐、刘梅杰
参赛单位：山东建筑大学
指导老师：薛一冰、崔艳秋

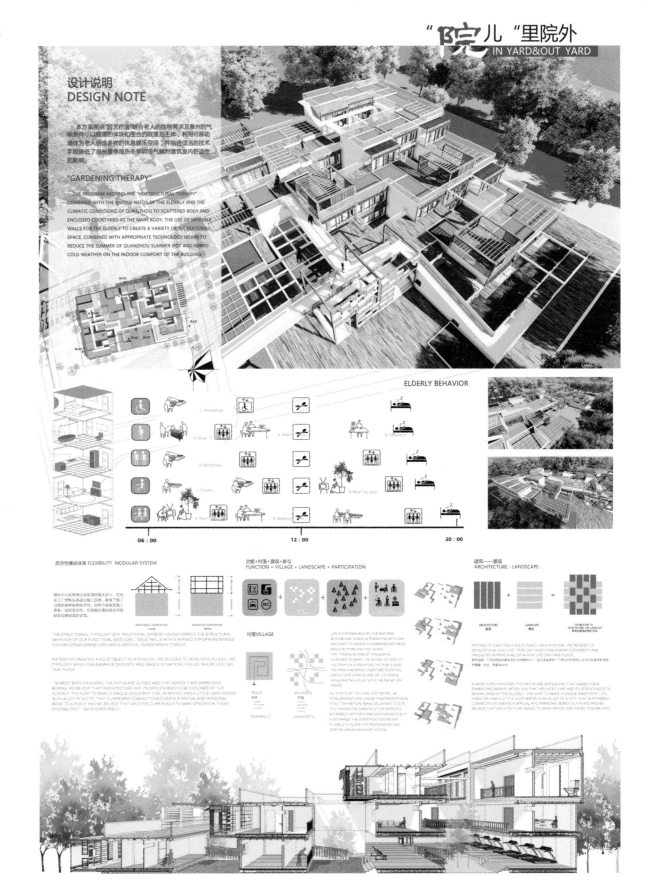

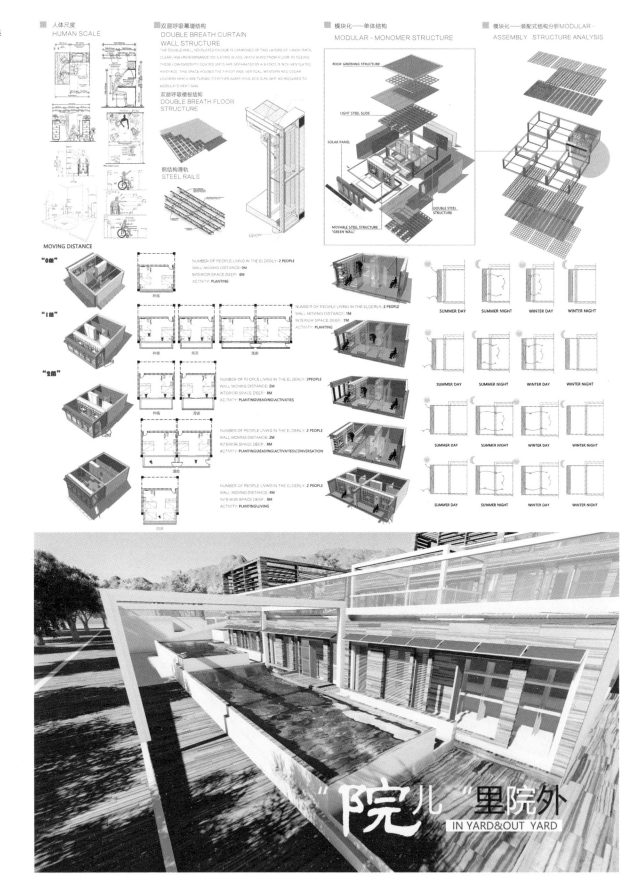

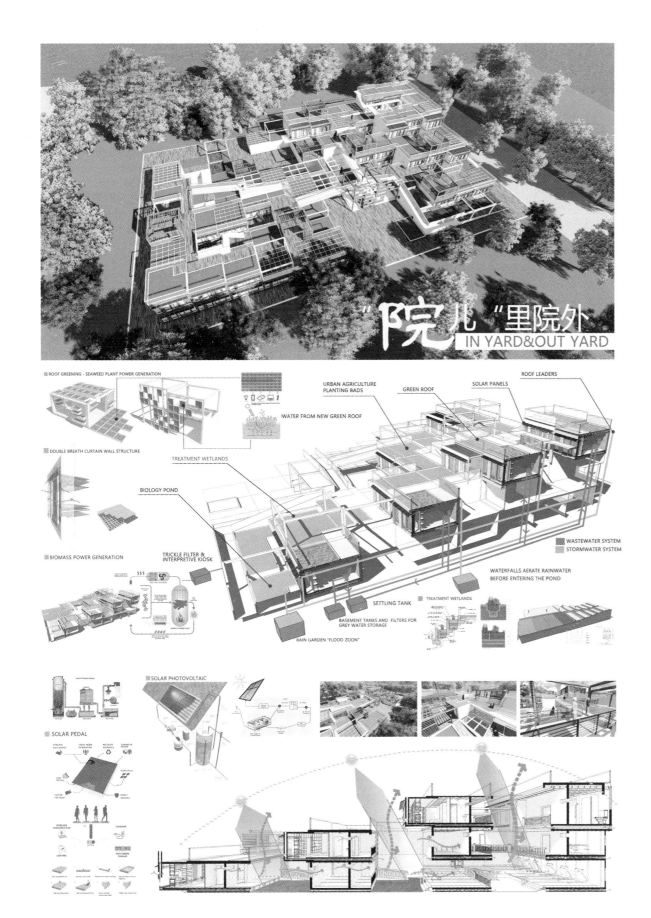

综合奖·优秀奖
General Prize Awarded · Honorable Mention Prize

注 册 号：4867
项 目 名 称：禅净·颐养服务中心（泉州）
Calm · Down Maintenance Service Center（Quanzhou）
作　　者：赵子豪、王娇婧、刘　畅
参赛单位：河北工业大学
指导老师：舒　平、岳晓鹏

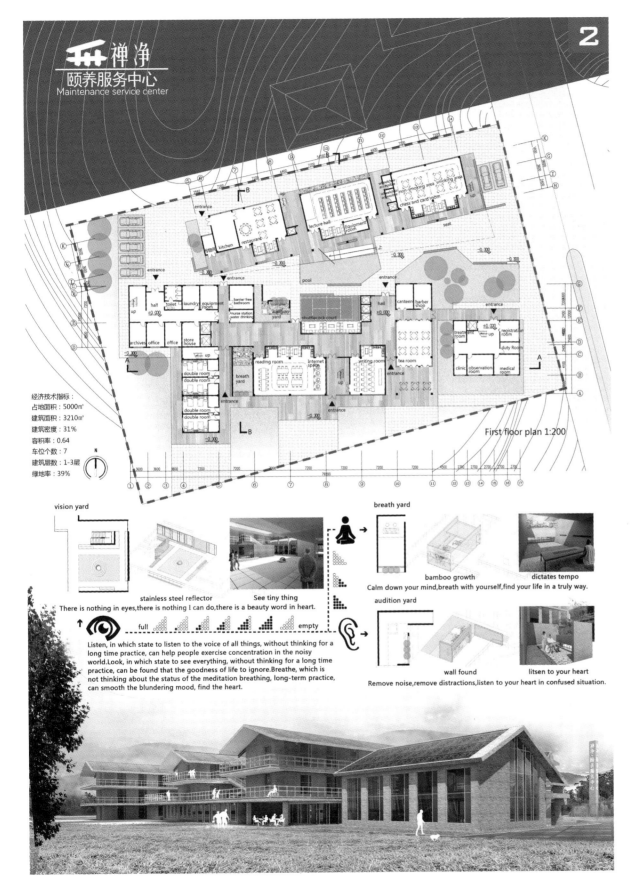

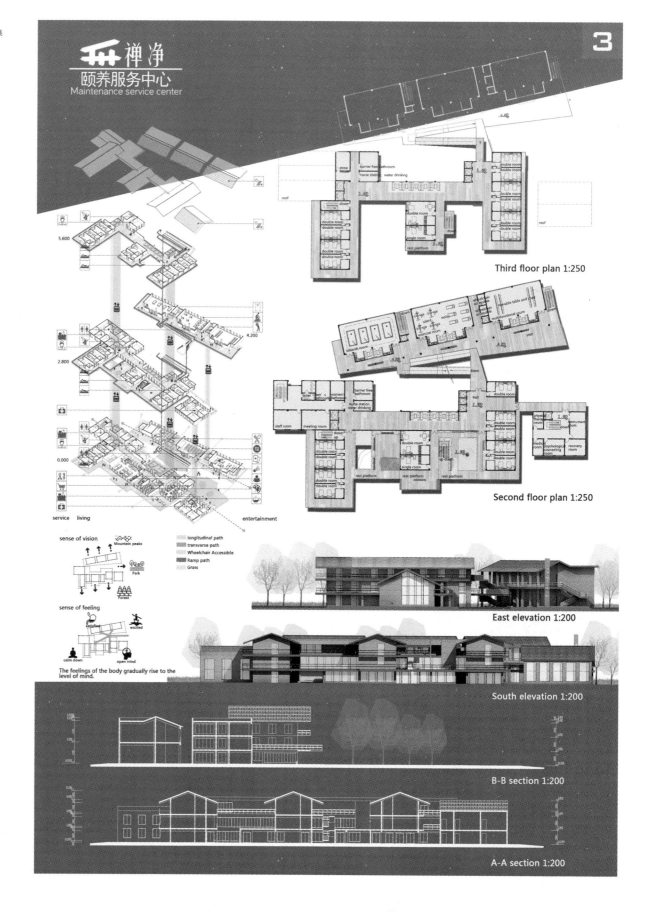

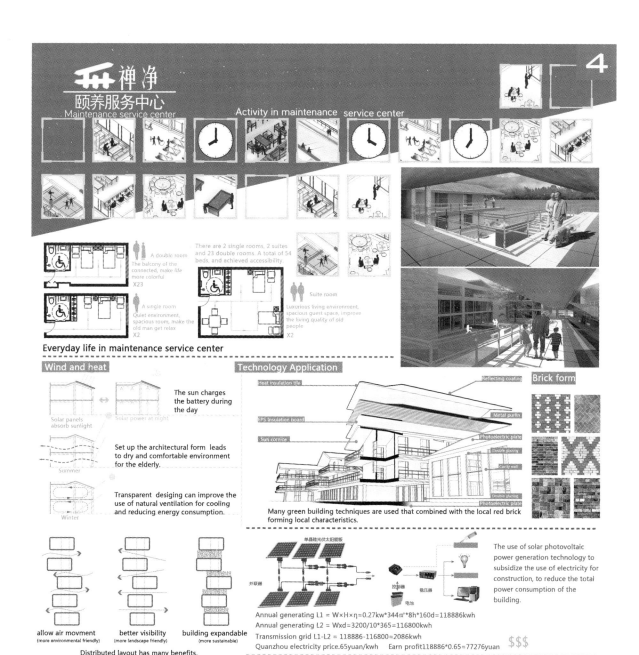

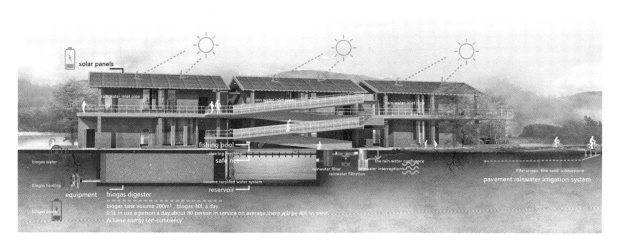

综合奖·优秀奖
General Prize Awarded · Honorable Mention Prize

注 册 号：4889
项目名称：记忆·寮院（泉州）
　　　　　Sunshine Yard of Memory (Quanzhou)
作　　者：景诗超、于　露、袁　梦、孙　慧
参赛单位：山东建筑大学
指导老师：张　勤

记忆·寮院 The Memory of Liao Yuan

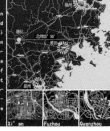
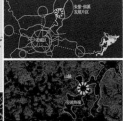

Base profile:
Quanzhou is located in Fujian Province, the base is located in the purple purple - Jiao Xi development area. It can be seen from the topographic map, because of the special climate impact, the building texture of Quanzhou is different from that of Xi'an area. The base is located in low-lying areas, surrounded by mountains, more prone to waterlogging.

设计说明：
　　该颐养中心设计以泉州地域特色建筑"手巾寮"为方案构思灵感源泉，总体布局形成开敞式入口广场以及仅供老人们使用的半私密的生活广场，二者遥相呼应，建筑内部设置两处庭院，并通过内廊使"通且敞"的寮院精髓在该设计中体现的淋漓尽致，达到建筑通风降温和除湿的效果。绿色节能技术方面采取"被动优先，主动结合"策略，被动策略通过布置生活小院、内庭院以及冷巷原理的运用，使建筑形成自身小气候进而达到通风采光降温隔热；主动方面采取立面种植绿化、鱼菜共生系统以及地道风结合官府发电系统通风遮阳板系统等。

Design description:
　　The remaining center in quanzhou region characteristic architectural design "towel" as the design inspiration, the overall layout form of type opening open entrance plaza and is only for old people to use half a private life square, both from afar building set up two courtyards, and through the gallery, make "and open" Lao yuan essence in the design reflect incisively and vividly, to achieve the effect of building ventilation cooling and dehumidification. Green energy-saving technology adopt "passive priority, active combination" strategy, passive strategy through life yard, in the courtyard and the use of principle of cold lane, cooling the building form its own microclimate and achieve ventilated daylighting heat insulation; Taken active facade planting green, fish food symbiotic system and tunnel ventilation visor wind combined with government power generation system, etc.

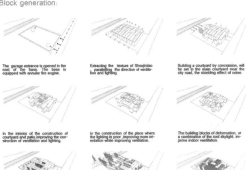

Block generation:

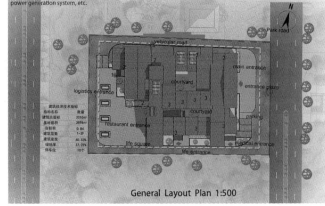

General Layout Plan 1:500

记忆·寮院——阳光下的疗养小院
The Memory of Liao Yuan—Nursing Courtyard under the Sun

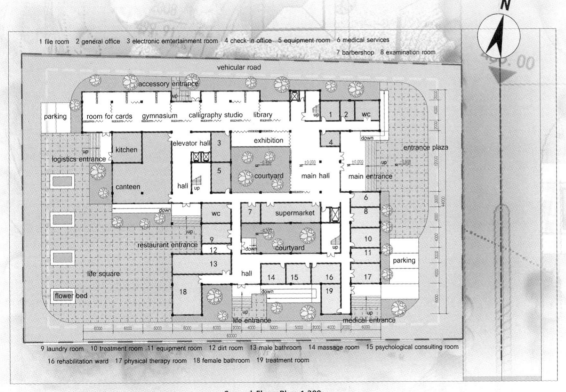

1 file room 2 general office 3 electronic emtertainment room 4 check-in office 5 equipment room 6 medical services 7 barbershop 8 examination room

9 laundry room 10 treatment room 11 equipment room 12 dirt room 13 male bathroom 14 massage room 15 psychological consulting room 16 rehabilitation ward 17 physical therapy room 18 female bathroom 19 treatment room

Second Floor Plan 1:200

South Elevation 1:200

记忆·寮院

The Memory of Liao Yuan

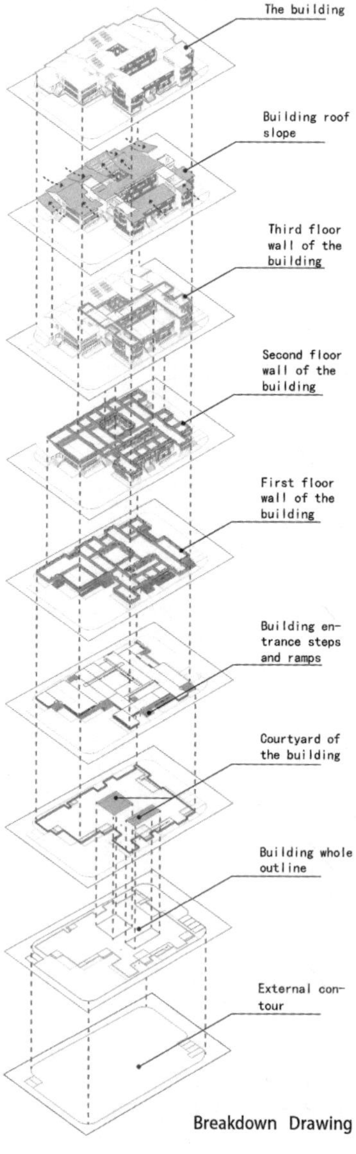
The building
Building roof slope
Third floor wall of the building
Second floor wall of the building
First floor wall of the building
Building entrance steps and ramps
Courtyard of the building
Building whole outline
External contour

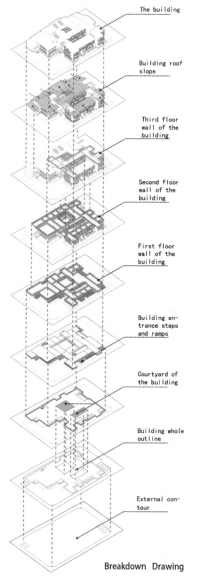

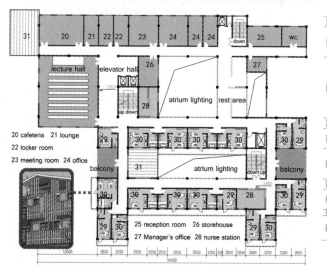

Breakdown Drawing

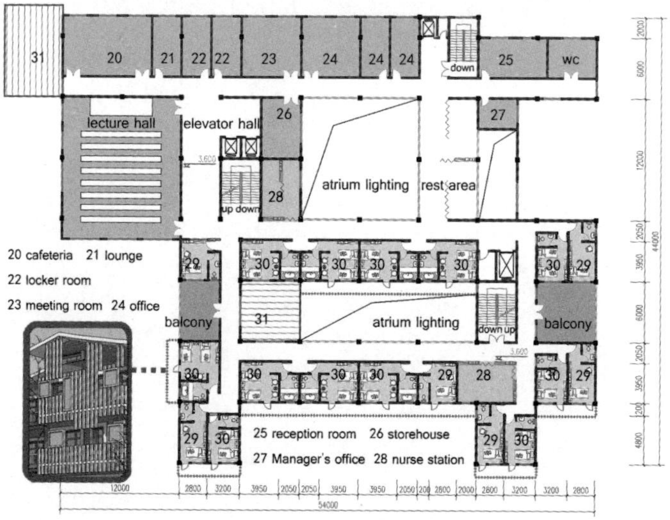
lecture hall
elevator hall
atrium lighting
rest area
balcony
down
up down
20 cafeteria 21 lounge
22 locker room
23 meeting room 24 office
25 reception room 26 storehouse
27 Manager's office 28 nurse station

Second Floor Plan 1:200

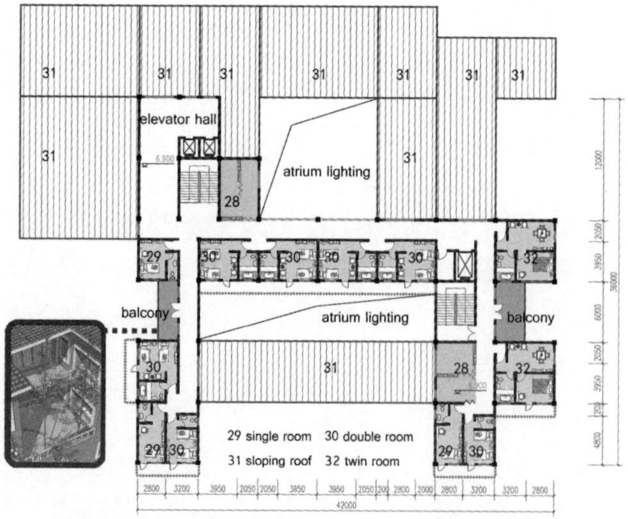
elevator hall
atrium lighting
balcony
29 single room 30 double room
31 sloping roof 32 twin room

Three Floor Plan 1:200

记忆·寮院——阳光下的疗养小院
The Memory of Liao Yuan—Nursing Courtyard under the Sun

APPLICATION OF PASSIVE TECHNOLOGY

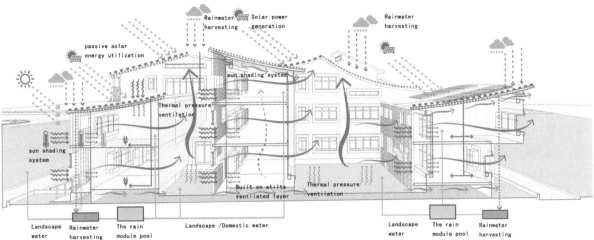

PASSIVE VENTILATION COOLING

1. Ventilation roofs

Winter night, the window is closed, forms the insulation cavity.

Summer, the window open, forming insulation cavity.

2. Double facade

Using the adjustable double facade shading lighting, heating, and different seasons.

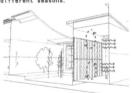 

Winter night, the window is closed, forms the insulation cavity.

Summer, the window open, forming insulation cavity.

Passive shading and heating

Using the adjustable double facade shading lighting, heating, and different seasons. Lateral elevation position set on the balcony ventilation, shading, solar energy integration vertical visor, and ACTS as a wind deflector, introducing natural wind indoor.

Winter night shutter closes, absorb heat during the day to the interior.

Summer night shutter fold, forming a coherent channel, ventilation and heat

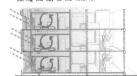 

Winter is colder days comes up hundreds of pages, form a coherent channel, daylighting heating.

Summer day, shutter rotation Angle, shading and ventilation.

3. Cold lane ventilation cooling

Cold lane summer cool during the day: the day of cold lane air is heated, the lower the air is relatively low, cold air trapped in the lower part, avoid heat into the interior. Cold lane summer night wind cooling: pull the night cold lane upper air cooling, lower relatively hot air, hot air rising heat can be taken away. Summer night cold lane storage: night temperature is low, oyster shell wall heat losses after the cold storage, indoor transmission cold during the day.

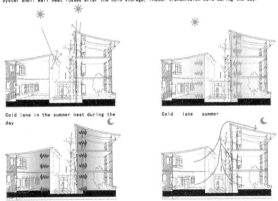

Cold lane in the summer heat during the day

Cold lane summer

Cold lane summer

Cold summer night

THE WEST FACADE PLANTING GREEN + FISH FOOD SYMBIOTIC SYSTEM

The west facade greening plants:
Combination of roof rainwater collection, wall landscape water environment, the construction of the western wall adopts modular vertical greening, effective utilization of rainwater greenery, excess water supply pool, pool water evaporation at the same time, improve the plant growth environment humidity in the air.

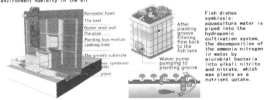

Fish dishes symbiosis: aquaculture water is piped into the hydroponic cultivation system, the decomposition of the ammonia nitrogen in water by microbial bacteria into alkali nitrite and nitrate, which was plants as a nutrient uptake.

WIND TUNNEL COMBINATION OF PHOTOVOLTAIC POWER GENERATION SYSTEMS, VENTILATION VISOR:

Wind tunnel system: wet hot air cooling after cooling tunnels, water vapor condenses from the air, dry cold air has been sent to the interior. Photovoltaic power generation system combined with wind tunnel system, bear the mechanical equipment such as fan power consumption; Ventilation visor ACTS as the solar chimney, heated hot air pollution will be emitted, adopt open wind tunnel design, provides plenty of fresh air for indoor.

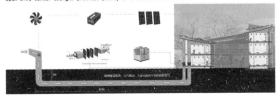

B-B Profile 1:200

记忆·寮院——阳光下的疗养小院
The Memory of Liao Yuan—Nursing Courtyard under the Sun

SPACE CHANGE ANALYSIS

Building a layer from the entrance hall to the southwest of the courtyard, north of space as can change activity space, mainly consists of folding partition space partition space with activities that have noise interference. Old people can according to the different time, different content, change the activity space. With limited land, the new, as far as possible to meet the old man use efficiency and amateur life.

Large social activities

Painting and calligraphy exhibitions

Pension preschool education joint activities

Condolences to the party

Movable wall collection form

Dining-room mobile space

APPLICATION OF SPACE SYNTAX

In some institutions endowment facilities, the same corridor and exactly the same as the room is very monotonous, the same scene appeared again and again will make a person feel very confusion, chaos. Space agency method can be more reasonable design of the application of endowment construction, guide the path design, monitoring set, stimulate the point set, etc.

Vision problems in architectural space, through the subdivision of the grid to see the line of sight depth, to explore the space for the line of sight of limited role.
Visual Step the Depth: from the selection set look outwards (yellow highlighted), for any other element in the global, passes through several view Depth. Guide nursing station, the layout of monitoring functions such as monitoring
line of sight with the old housing hallways, elevators, the relationship between public space

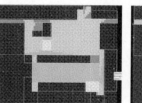

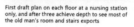

First draft plan on each floor at a nursing station only, and after three achieve depth to see most of the old man's room and stairs exports

Scheme revised set two nursing station, station to two exit of accommodation area, most of the old man has a depth value room, which can be immediately dynamically observed the old man

Visual Integration (HH) : a global view the most shallow depth, the easiest to see where the Clustering Coefficient: space boundary limit effect in Visual/elements on the strength of the camouflage effect. (assuming that the space of the border are opaque entity interface). Considering the function layout problem.
Red area is better area of illicit close sex, what kind of behavior in this space? Cameras monitoring approach?

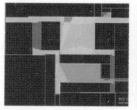
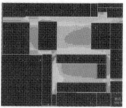

West Elevation 1:200   East Elevation 1:200

记忆·寮院——阳光下的疗养小院
The Memory of Liao Yuan—Nursing Courtyard under the Sun

A-A Profile 1:200

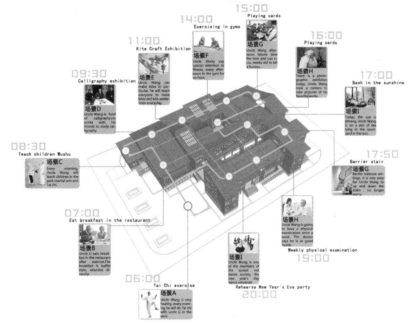

TO ADAPT TO THE CLIMATE CHARACTERISTICS OF TECHNICAL STRATEGY

Dehua is a subtropical climate, high temperature and rainy summer, winter gentle little rain, with a moderate temperature cool, clear four seasons, abundant rainfall, rain heat same characteristics. Summer mainly ventilation cooling, shading and cooling, ventilation, dehumidification, both passive heating in winter.

Climate characteristics and design requirements   Technology strategy

综合奖·优秀奖
General Prize Awarded · Honorable Mention Prize

注 册 号：4958
项目名称：井·巷（泉州）
　　　　　Well & Lane（Quanzhou）
作　　者：岑士沛、傅嘉伦、冯嘉瑜、
　　　　　张渡也、朱琪琪
参赛单位：深圳大学
指导老师：艾志刚

井·巷
WELL & LANE
泉州生态颐养服务中心项目设计
可再生绿色能源的归宿

Location analysis

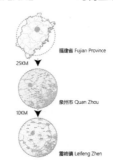

The base is located in the city of QuanZhou where is the famous capital of ceramic, at the same time, it assemble ecological habitat, spa vacation, sports and leisure, business meetings, star hotels as one of the old-age health resort town. The full use of solar energy, other renewable energy technologies and peripheral superior natural environment will be the our fundamental design points for the subtropical climate zone in the green, low-carbon, healthy ecological service center.

设计说明 Design Description

The case is located in Quanzhou, Fujian Province, a subtropical monsoon climate, high temperature and rainy, so the case at the beginning of the design will take full account of the cooling ventilation and drainage, while making full use of solar energy resources. The study of Quanzhou local traditional houses, we choose "hand towel Liu" form of learning, residential part of the short bay into the deep, each floor has a number of erroneous patio to promote ventilation, while enriching the elderly living space. According to the base and the surrounding environment analysis, the base of the right side of the environment for public activities, while other services for the park the elderly. The left side of the base with the scattered layout of the way the elderly housing, single-gallery and the bottom of the overhead makes the activities of space ventilation effect is very good, rich communication space. The sloping roof of the residential unit is not only conducive to rainwater harvesting, but the 15 degree slope roof is best suited for placement of solar panels.

Base And Target Analysis

Base located within the eco-home, residential planning surrounding the massive pension. Task request of newly-built homes in public with other live parts to the Park Service. So according to the relationship between space and environment, we will be building entrances and external activities are set at the right side of the base, open posture in the elderly in hospital.

Climate-Analysis Of Quanzhou

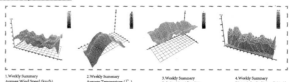

1. Weekly Summary Average Wind Speed (km/h)　2. Weekly Summary Average Temperature (℃)　3. Weekly Summary Relative Humidity (%)　4. Weekly Summary Direct Solar Radiation (W/m2)

Southern region is located in a subtropical hot and humid monsoon climate with, in our climate zone In summer hot and winter warm area, summer heat is essential. Annual average temperature is above 10 degrees Celsius along the coast. Warm in winter, January mean temperature of 7~10 c in coastal areas, mountains 6 - ℃. Summer. Hot and mean temperature of 20 -39 c. 1 400 -2 000 mm of annual precipitation, annual sunshine duration is 1 700 -2 300 h, large quantity of solar radiation.

3-Prevaling Winds-Wind Frequency (Hrs)
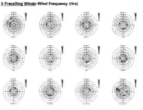

4-Optimum Orientation

Taiwanese traditional dwellings on the site of the first is to consider building towards, reasonable towards seeking a favorable to make full use of natural lighting and ventilation, while avoiding adverse toward excessive sunlight, thereby reducing the energy consumption of the building.

6-Comfort Percentages

The rich wind resources in Quanzhou, summer Average wind speed in the 2.9m/s, as long as the reserved channel during the architectural design, you canAchieve the purpose of ventilation and dehumidification

Calculation of energy-saving software technology, we can design, passive solar heating, natural ventilation, night-time ventilation, evaporative cooling, indirect evaporative cooling on the great progress. Greater in multiple complex passive technology improved.

Environment Functional Analysis

New homes located in the porcelain impressions Eco-Park, is a ecological Habitat, spa resorts, sports and leisure, business meetings, star hotel as one of the old-age and health resort towns. We need to cater to the surrounding atmosphere to set the function and shape of the building.

The Quote of "Shoujinliao" Traditional Dwelling

One of the traditional houses in Quanzhou, the name "towel hut"("手巾寮"), refers to the slender like a towel, more for ordinary people's homes. Bar-single family residential buildings, built along the streets, and appear as a dense row groups, creating a variety of street space. Homes were single-Bay, width 3M-4m, layers of deep inward to the center of 20m. "The towel hut" ("手巾寮") houses after a long period of evolution, has formed a completely system of lighting, ventilation and heat-resistant construction.

井·巷
WELL & LANE — 可再生绿色能源的归宿
泉州生态颐养服务中心项目设计

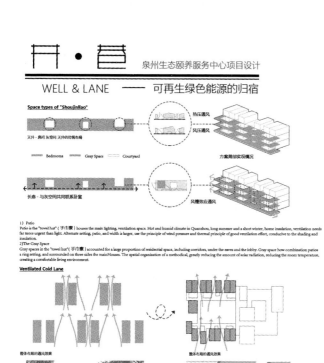

1) Patio
Patio is the "towel hut" (手巾寮) houses the main lighting, ventilation space. Hot and humid climate in Quanzhou, long summer and a short winter, home insulation, ventilation needs far more urgent than light. Alternate setting, patio, and width is larger, use the principle of wind pressure and thermal principle of good ventilation effect, conducive to the shading and insulation.

2) The Gray Space
Gray spaces in the "towel hut" (手巾寮) accounted for a large proportion of residential space, including corridors, under the eaves and the lobby. Gray space how combination patios a ring setting, and surrounded on three sides the mainHouses. The spatial organization of a methodical, greatly reducing the amount of solar radiation, reducing the room temperature, creating a comfortable living environment.

Ventilated Cold Lane

(3) Laneway
Lane Gallery is "the towel hut" (手巾寮) residential traffic spaces before and after contact, width smaller throughout, making it a "shroud" (风槽) formed shroud effect, namely wind blowing from a wide range of outdoor into a narrow channel, wind speeds increased, driving air around, form a good general ventilation. Meanwhile, lane Gallery short of sunshine, cool temperatures. Such a double combination, forming a pleasant lane gallery space.

5-The origin of the yard
The origin of the yard

多方案对比 Schemes Comparison

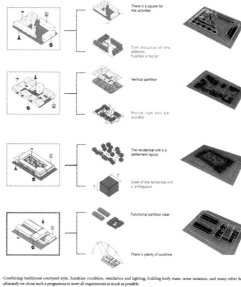

Combining traditional courtyard style, Sunshine condition, ventilation and lighting, building body mass, noise isolation, and many other factors, ultimately we chose such a programme to meet all requirements as much as possible.

summer ventilation strategy is focused on cross ventilation. pressure differences encourages cross ventilation also cools the building down.

winter ventilation strategy is focused on heat recycling with limited (to around 20% of overall air cycle) incoming fresh air. In addition, indoor air it is being purified and humidified by plans in a conservatory.

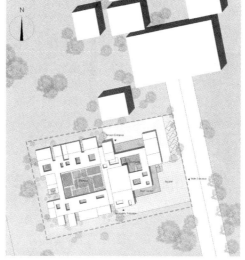

总平面图 1:500

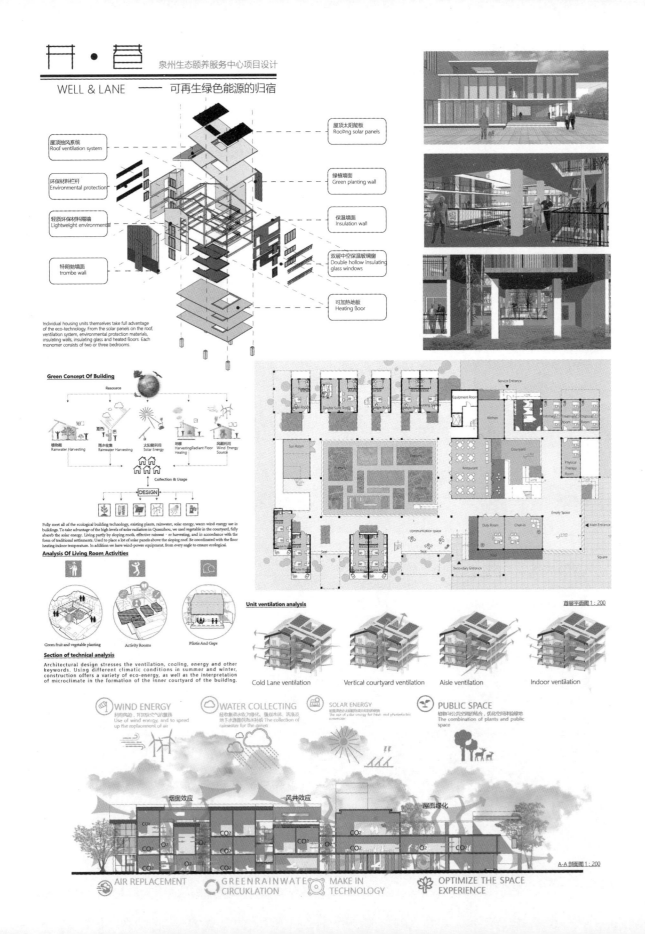

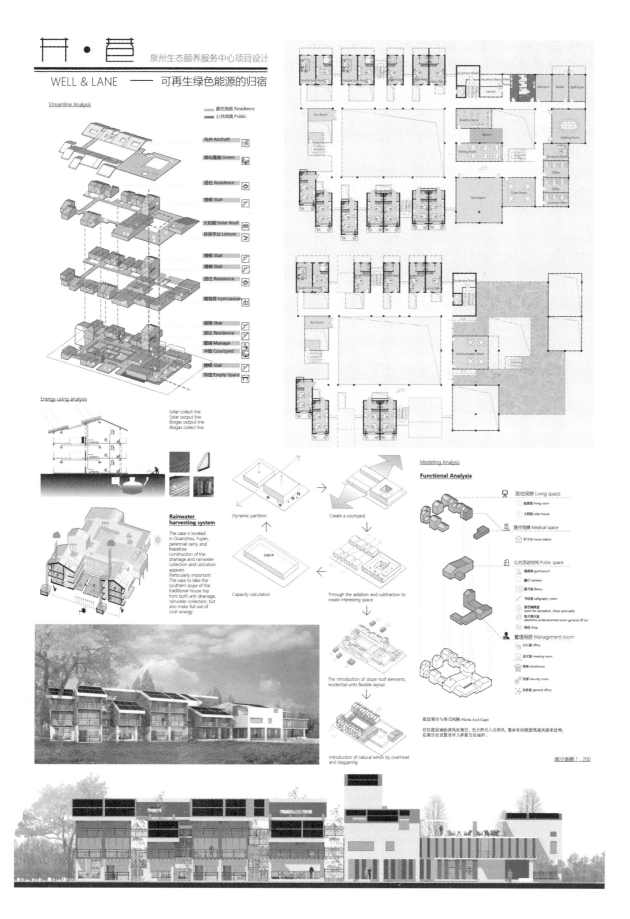

综合奖·优秀奖
General Prize Awarded·
Honorable Mention Prize

注 册 号：5011
项目名称：光·转·折（泉州）
　　　　　Light·Turn·Fold (Quanzhou)
作　　者：潘高、邹明妍
参赛单位：重庆大学
指导老师：周铁军、杜晓宇、张海滨、
　　　　　Regina Bokel（Delft
　　　　　University of Technology）

光·转·折 LIGHT·TURN·FOLD

泉州生态颐养服务中心项目设计

SITE ANALYSIS

Location Analysis

Base Analysis

The Base Path Analysis　Function Surrounding the Base　Base of landscape Element Analysis　Building Facade Surrounding the Base

DESIGN DESCRIPTION

设计说明：
　　基地位于福建省泉州市德化县雷峰镇瓷都印象生态园内，北纬25°56′，东经118°32′，地处福建省中部，泉州西北部。紧邻203省道，距德化县城7公里。基地用地平整，周围环境优美。本次设计是为老人使用的生态颐养服务中心。
　　老人是一个特殊的群体，需要得到我们更多的关注和呵护。本方案以"老人、阳光、节能"为中心设计思想，充分考虑老人的身体和心理需求，以泉州传统民居空间为原型，进行太阳能技术的创新设计，为老人提供一个自然、舒适、节能的颐养空间。

　　Site is located in Quanzhou City, Fujian Province. Base is located in central Fujian Province, Quanzhou northwest. Adjacent to 203 provincial highway, 7seven kilometers away from Dehua county. Dehua County is located in the roof of the Dai Yunshan District. The base of land is flat, surrounding a beautiful environment. This design is the ecological service center care for the old.
　　The old man is a special group, who need our more attention and care. This scheme centered on "the old , sunlight, energy saving" as the design idea, give full consideration to the old man's physical and psychological needs. What's more, use the space of traditional houses in quanzhou as the prototype, for solar energy technology innovation design. Finally, to provide a natural, comfortable, energy saving of the remaining space for the old man.

PSYCHOLOGICAL NEED OF THE OLD

Through the analysis of the space of traditional houses in quanzhou, elicit the following four types of space to be suitable for the old

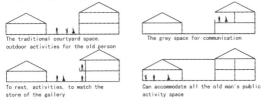

Problems----Suit the remedy to the case
1、The sense of belonging----Find the memory of the space, can let buildings to rediscover their own characteristics, let building real sense of place, let the old man found himself the most real sense of belonging and identity.
2、Loneliness----Don't have my children for a long time, the old man more need to accompany, this design in meet the demand of old people living at the same time, offers a wide range of all kinds of public space, meet the needs of old people of different types.
3、Light sensation----Abundant sunshine make the old man's physical and mental healthy, postpone building texture, building towards is not perfect, to reverse direction of the window, and solar energy integration design, can make the room to receive more sunlight.

REGIONAL ARCHITECTURE

Quanzhou Ecological Remaining Service Center Project Design

光·转·折

光·转·折 LIGHT·TURN·FOLD

■ CLIMATIC CONDITIONS ANALYSIS

■ Wind Environment Analysis — Wind Frequency

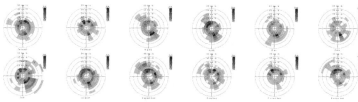

The wind frequency chart of the 12 month on the left side directly shows the wind direction and wind speed in Quanzhou area. The northeast wind is mainly in winter, the southeast wind and the northwest wind is mainly in summer and the wind speed in the summer is small, relatively humidity in Quanzhou is large, we need to consider the use of natural ventilation to achieve the purpose of dehumidification.

■ Wind Environment Analysis — Wind Temperature

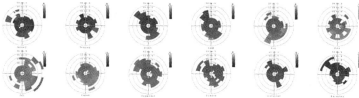

The wind temperature map chart of the twelve month on the left side directly shows the wind temperature in Quanzhou area, the wind temperature greatly affect the comfort of the environment, the temperature of other transitional season which is April, May, October, November is more comfortable.

■ Enthalpy Diagram Analysis

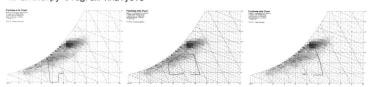

This figure shows the impact of passive solar heating, high heat capacity materials, natural ventilation on the comfort. The natural ventilation and the use of high heat capacity of the material have a great impact on comfort. However, passive solar heating has little help on the comfort.

■ Daily Average Temperature

■ Relative Humidity

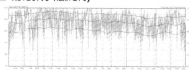

The temperature fluctuation in Quanzhou is relatively small, and the peak temperature which appears in June, July, August is higher than that of human body. The temperature which is in December, January, February is lower than that of human body. The relative humidity is large and the change of the relative humidity is not obvious.

■ Annual Solar Radiation Analysis

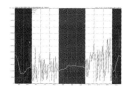

■ Best Orientation Analysis

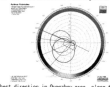

The best direction in Quanzhou area close to 175° 202.5°

■ SITE PLAN 1:500 / OVERALL ANALYSIS

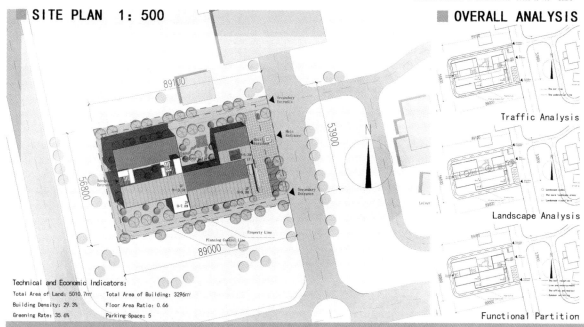

Technical and Economic Indicators:
- Total Area of Land: 5010.7㎡
- Total Area of Building: 3296㎡
- Building Density: 29.3%
- Floor Area Ratio: 0.66
- Greening Rate: 35.6%
- Parking Space: 5

Traffic Analysis
Landscape Analysis
Functional Partition

Quanzhou Ecological Remaining Service Center Project Design

光·转·折 LIGHT·TURN·FOLD

03

泉州生态颐养服务中心项目设计

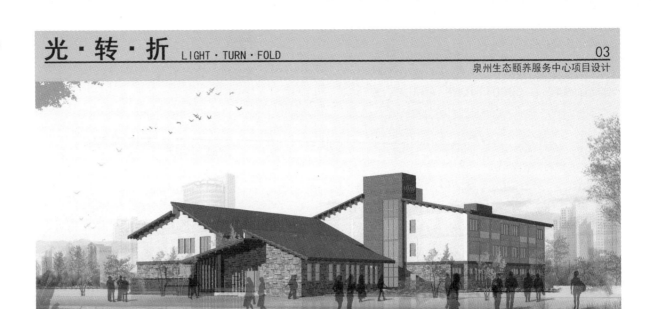

INSOLATION ANALYSIS

Insolation Sunshine in Summer

Insolation Sunshine in Winter

FIRST FLOOR PLAN 1:200

Insolation Time Analysis in building plan

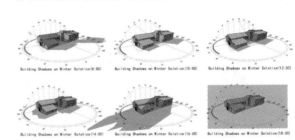

Quanzhou Ecological Remaining Service Center Project Design

光·转·折

光·转·折 LIGHT·TURN·FOLD

泉州生态颐养服务中心项目设计

SECOND FLOOR PLAN 1:200

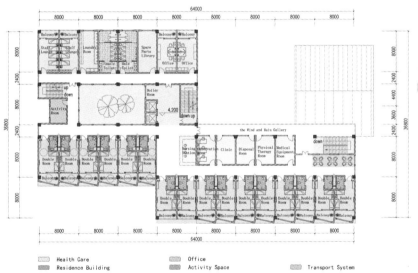

- Health Care
- Residence Building
- Office
- Activity Space
- Transport System

BLACONY ORIENTATION ANALYSIS

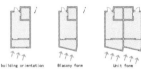

VARIABLE SPACE

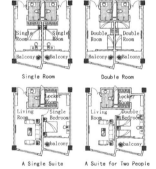

ELEVATION

East Elevation 1:200   Wast Elevation 1:200

Quanzhou Ecological Remaining Service Center Project Design

光·转·折

光·转·折 LIGHT·TURN·FOLD

05

泉州生态颐养服务中心项目设计

SOLAR RADIATION ANALYSIS

Using the ecotect software, the four elevations and the tops of a cube were analyzed by monthly solar radiation illumination on a monthly basis. The data are compared and analyzed.

Under the same area, the solar radiation intensity of the roof was more than four facades, the west facade of the solar radiation in turn higher than the other three facades. In this design, we need to take measures such as planting walls to reduce the solar radiation. We use solar panels to collect energy on roof and south side.

Roof

South Facade

North Facade

East Facade

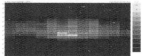
West Facade

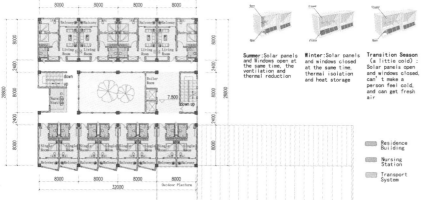

THIRD FLOOR PLAN 1:200

WINDOW ANALYSIS

Summer: Solar panels and Windows open at the same time, the ventilation and thermal reduction

Winter: Solar panels and windows closed at the same time, thermal isolation and heat storage

Transition Season (a little cold): Solar panels open and windows closed, can't make a person feel cold, and can get fresh air

- Residence Building
- Nursing Station
- Transport System

THERMAL ANALYSIS

Summer Day: South ventilation valve closed, cooler air from the north into the double floorlayer, reduce indoor temperature by radiation

Summer Night: Open all valves, cooler air form convection through the double-layer floor, and hot air out construction, so as to reduce indoor temperature

Winter Day: Sunlight to heat the south balcony air temperature, higher temperature of the air form convection through the north of double-layer floor, so as to improve the indoor temperature in winter

Winter Night: The balcony was heated during the day, form convection and exchange with indoor air at night, so as to improve the indoor temperature

VENTILATION ANALYSIS

Summer: In the south to the sun, the flow of south air is accelerated, so as to drive the internal air circulation of the whole building

Winter: Outdoor air temperature is low in winter, close the ventilation valve, air temperature between the south sun increase to promote the circulation of air indoors, so as to improve the indoor temperature

LIGHTING ANALYSIS

Summer: Angle of the sun is higher in summer, the roof overhangs can effectively keep out the sun's rays, and avoid too much sunlight into indoor, so as to reduce the indoor temperature in summer

Winter: Angle of the sun is lower in winter, the roof overhangs can't effectively keep out the sun's rays, and keep sunlight into indoor, so as to increase the indoor temperature in winter

South Elevataion 1:200

Quanzhou Ecological Remaining Service Center Project Design

光·转·折

光·转·折 LIGHT·TURN·FOLD

DAILY DEMAND OF SOLAR HOT WATER

Estimation of Daily Hot Water Consumption

In accordance with the relevant provisions of the "Tcivil building water-saving design standards" (GB 50555-2010). The hot water consumption of Per capita 60 ℃ is 40 L/d. The number of beds is 48 in this design. The average daily consumption of hot water in the living part is 40L/d × 48 =1920 L/d. Assuming the total number of people in the design of the public bath shower is 50 per day, the hot water consumption of Per capita 60 ℃ is 50 L/d, the average daily consumption of hot water in public shower part is 50 L/d × 50 =2500L/d.

Estimation of Water Consumption for Solar Hot Water Heating

The room temperature is calculated at 18℃ of November, December, January, February, March in winter. Heating heat load can be calculated according to 60W/m², Rough estimate, the total number of heating hot water of 60 ℃ is 30000L/d.
Estimation of Daily Hot Water Consumption, 2500L/d+30000L/d=32500L/d.

Calculation of Collector Heating Area

According to the formula, $Ac = Qw \cdot Cw \cdot (t_{end}-t_i) \cdot f / J_r \cdot \eta_{cd}(1 - \eta_L)$
- Ac——collector heating area (m²)
- Qw——daily water consumption (kg)
- Cw——specific heat capacity of water
- t_{end}——final temperature (℃)
- η_L——heat loss rate of storage tanks
- η_{cd}——collector efficiency
- t_i——initial temperature (℃)
- f——solar guarantee rate
- J_r——daily radiation (KJ/m²)

The required collector area is 460 m², the actual area of solar collectors is 474 m².

SUPPLY AND DEMAND OF SOLAR PHOTOVOLTAIC POWER GENERATION

The design of the electricity consumption is 80W/m² (included in air conditioning electricity, no air conditioning electricity can be reduced by 40% to 50%), with a total construction area of 3296 square meters. Quanzhou area effective daily sunshine time is 6 Hourly calculation, rough estimate.

Month	Electricity Demand (Wh)	Electricity Bill (yuan)	Power Generation (KW·h)	Electricity Generation Bill (yuan)
January	5000	3500	6279.458	3865.84
February	5000	3500	6458.283	3946.75
March	4800	3360	7581.616	4056.89
April	4600	3220	7601.614	4531.27
May	4600	3220	9247.467	4835.63
June	4600	3220	11278.394	5184.23
July	5000	3500	14371.643	5218.35
August	5000	3500	14030.581	5148.27
September	5000	3500	11394.443	4639.35
October	4600	3220	10558.793	4973.16
November	5000	3500	7857.359	4312.15
December	5000	3500	6693.549	4014.93
Total	58200	40740	113353.581	50669.93

TECHNOLOGY STRATEGY ANALYSIS

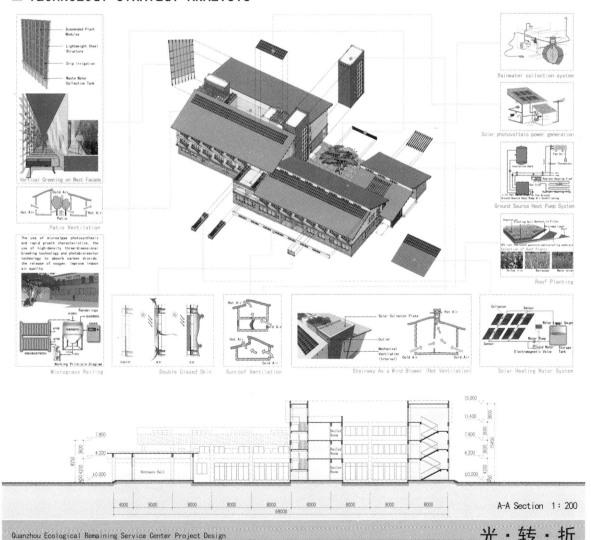

A-A Section 1:200

Quanzhou Ecological Remaining Service Center Project Design

综合奖·优秀奖
General Prize Awarded · Honorable Mention Prize

注 册 号：5144
项目名称：围院——太阳能养老院设计
（泉州）
Espace Elastique (Quanzhou)
作　　者：贾兆元、李倩芸、李　磊、
　　　　　凌　艺、郭梦铭、薛冰琳
参赛单位：北方工业大学
指导老师：马　欣、赵春喜

围院——太阳能养老院设计 1

设计说明
Design Specification

方案的灵感源于泉州地区常见的三合院民居的形式，整体是一个大院子，在院子中又有若干个小院子，形成院中有院的整体平面布局。屋顶的形式源于泉州传统民居高高翘起的飞燕脊形象，从各个角度看，屋顶层层叠叠错落有致，使人联想到泉州著名的古老建筑。在太阳能技术的使用上，充分利用太阳能被动技术，使得室内采光充足，夏季通风顺畅，辅助以太阳能主动技术，通过在屋顶上的太阳能光伏瓦将太阳能转化利用。以解决泉州地区冬季阴冷潮湿，夏季炎热的问题。

The inspiration of this plans stems from the living style of triple-house courtyard commonly seen in Beijing area. Generally, it is a lager courtyard with several small yards among it, which forms the overall plane layout of yard within yard. The form of rooftop derives from the image of Quanzhou traditional flying swallow ridge. The roofs of the houses overlap and form a patchwork pattern which reminds people of the ancient architecture. By adopting solar energy technology, the house enjoys a good daylighting and comfortable ventilation in summer, which has solved the problems of sullen winter and hot summer. Besides, with the aid of solar energy auto-technology, the solar energy can be transferred through the solar photovoltaic tiles.

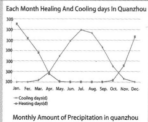

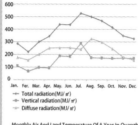

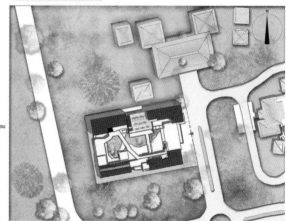

SITE PLAN 1:1000

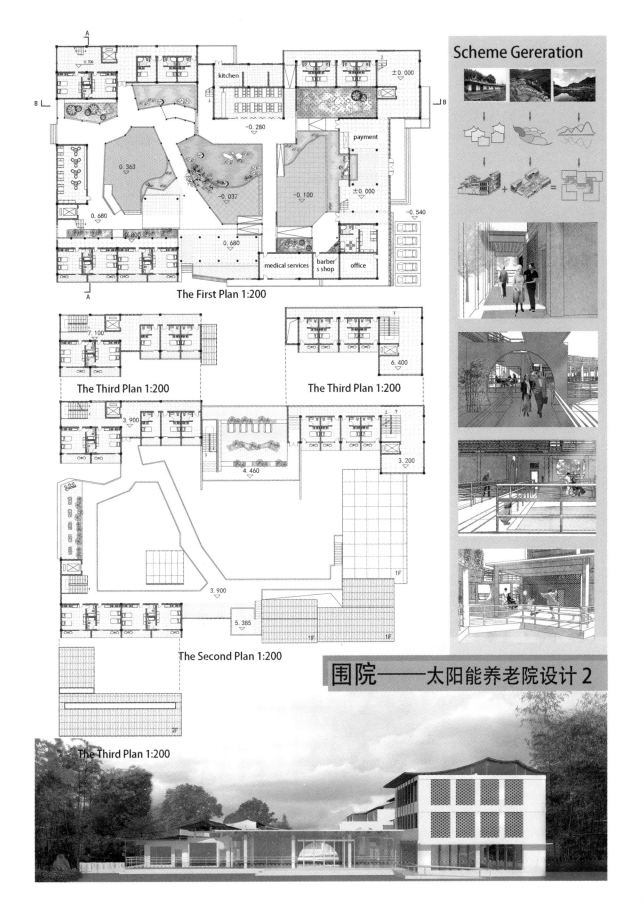

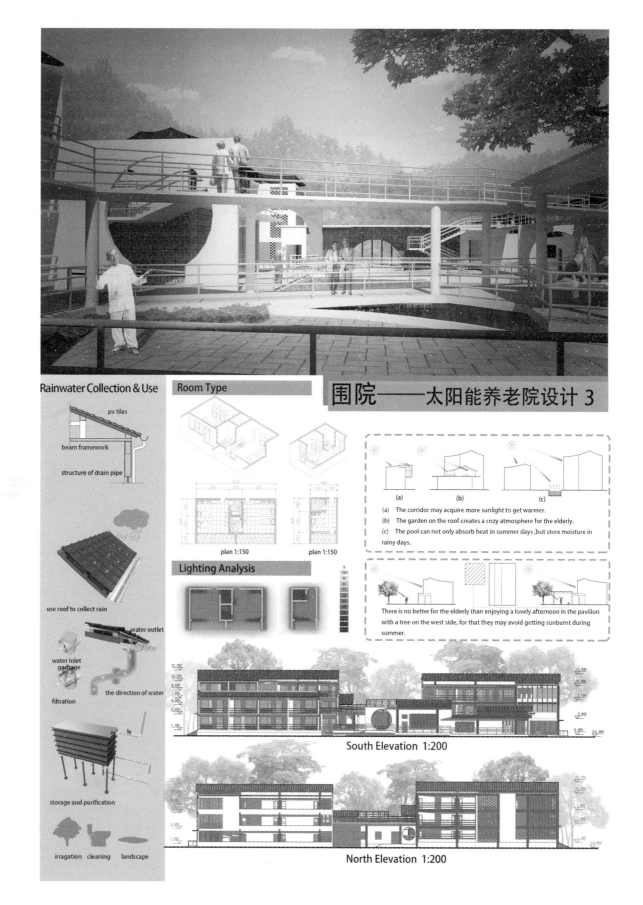

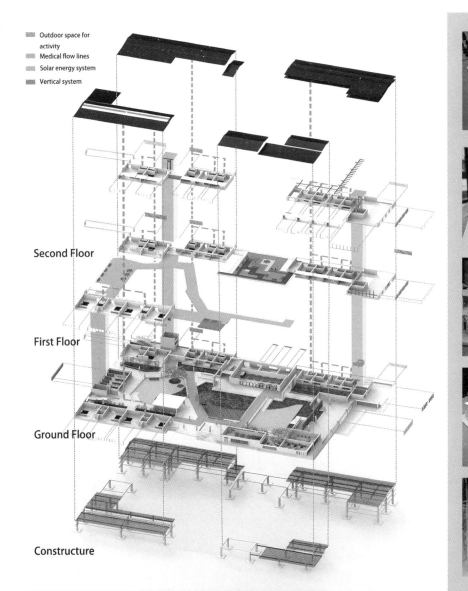

- Outdoor space for activity
- Medical flow lines
- Solar energy system
- Vertical system

Second Floor

First Floor

Ground Floor

Constructure

围院——太阳能养老院设计 5

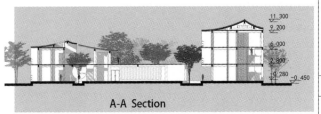

A-A Section

B-B Section

Economic And Technical Index

Order	Function			Num(room)	Area(㎡)
First floor	Residential living room	Single room		6	174
		Double room		7	210
	Living room	Living assistance room	Canteen	1	82.8
			Kitchen	1	60.7
			Communication hall	1	74.3
			Barber shop	1	26.5
	Management service room		storeroom	1	20
			duty room	1	20
			Reception room	1	140
			Office	1	37.5
			Staff rest room	1	19.8
			Staff living room	1	22
	Health care room		Treatment room	1	52
	Public activity area	Acivity room	Room for recreation, chess and cards	1	67.3
			Traffic area		800
Second floor	Residential living room	Single room		7	203
		Double room		6	234
	Public activity area		Traffic area		590
Third floor	Residential living room	Single room		3	72
		Double room		2	58
	Public activity area		Traffic area		190

Active Solar Energy Systems

Based on the previous research on the environment of the Quanzhou, we choose solar photovoltaic panels, lead-acid batteries, heating system and Low-E double glass as the main means of energy saving. The heat loss of the windows and doors is the main part of the building energy consumption. Using the doors and windows which are made of Low-E glass can greatly reduce the heat-losing caused by radiation from indoor to outdoor, achieving the ideal energy saving effect.

Compared with common tiles, the high-performance tiles which have the characteristics of Efficient insulation, strong waterproof, light weight and long service life, were made of clay, mixing with a variety of special materials. Through the combination of encapsulation technology and solar cells, we succeed in making the solar cells to generate electricity, with the preservation of original architectural style. Eventually, this kind of high-performance tiles which can generate electricity were invented.

Low-E Insulated Glass

Natural Ventilation

Floor Heating

External Windows Insulation System

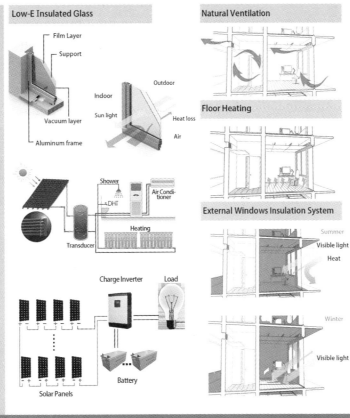

围院——太阳能养老院设计 6

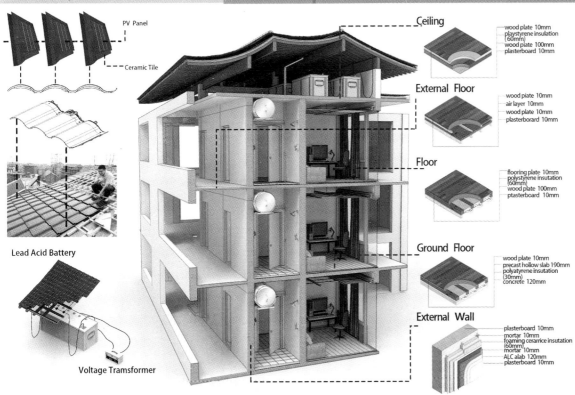

综合奖·优秀奖
General Prize Awarded · Honorable Mention Prize

注 册 号：5229
项目名称：游廊串"绿"（泉州）
　　　　　Play in the Green Corridor
　　　　　（Quanzhou）
作　　者：李　磊、主曼婷、吴鹏龙、
　　　　　刘姝伶、刘义鹏
参赛单位：合肥学院、合肥工业大学
指导老师：李　娟、胡　毅、谢雪胜

游廊串"绿" Play in the "Green" Corridor
泉州养老院设计 1

文字说明/Illustration

以自然中的"绿"色——光、水、风三种元素为主题进行院落的植入，达到空间的升华。以"绿廊"为媒介，使院落交融渗透的同时联系着老人的活动与院落空间。本案坐落在瓷都印象生态园里面，四周群山环绕，景色优美，建筑采用传统的坡屋顶，以增加老人与建筑之间的亲和力，建筑与山之间的呼应，整个建筑采用了合院式布局，增加了老年人们间的联系。

Use the three "green" elements in the natural-light, water and wind in order to implant courtyard so that the space can be sublimed. Though the medium of "Green corridor", the courtyards blend and the aged actives keep connection with courtyard space. The case with beautiful sceneries is located in the porcelain city ecology, surrounded by mountains. The building use sloping roof to increase appetency between the aged and space, building and mountain in cooperation with each other. The courtyard layout is adopted in the whole building, increase the aged connection with each other.

▎Site Analysis

Quanzhou, Fujian　　Dehua　　project site selection

▎Status Analysis

The surrounding environment is beautiful | The surrounding transportation is convenient | The surrounding landscape resources are abundant | Be situated in the Hot summer and Warm winter area

▎Design Thinking

step1: Think about what building resonates with the aged

step2: Using local porcelain firing into red brick cover and sloping roof to cater to the local culture

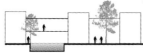

step3: Increase the courtyard and corridor space raise the contact among the aged

step4: Use of solar energy technology to provide comfortable living conditions for the aged

▎Base Analysis

According to the axis to locate main entrance | According to the surrounding buildings cutting itsel | According to the surrounding buildings layout implanting the inner courtyard | According to the surrounding buildings using the sloping roof

▎Block Analysis

Regular crustal block | Cater to terrian | Cut block | Form unit

Permutate and combinate | Increase corridor space | Implant courtyard | Implant sloping roof

游廊串"绿" Play in the "Green" Corridor 泉州养老院设计 2

Site area: 4925.7m²
Building area: 3278.5m²
Floor area: 1683.5m²
Greening rate: 37%
Floor area ratio: 0.67
Number of parking Spaces: 7

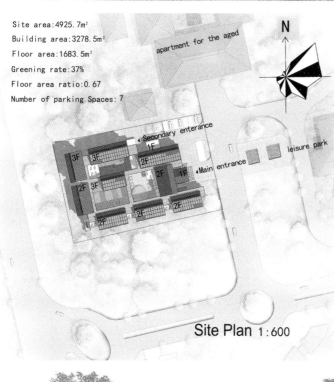

Site Plan 1:600

Courtyard Analysis

狭长的水院空间给人以静谧感，适合老人喝茶、思考、回忆。
Long and narrow water courtyard make people feel quiet, fit elder drink tea, tink and recall.

舒展的风院空间，给人以开阔清爽的感觉，适合老年人锻炼、散步、娱乐。
The wind courtyard is extended which can give people wide and refreshing feeling, fit elder go for a walk and do sports.

亲切的光院空间，光影透过绿植的缝隙洒下，给人以温暖、惬意之感，适合老年人交流。
The light courtyard space is amiable. Light shed through the plants which make people feel warm and pleased, fit elder communicate.

Action Analysis

Before go to Nursing Home

CONTRAST

In Nursing Home

morning

ackemma

noon

afternoon

dusk

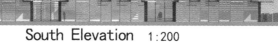

South Elevation 1:200

游廊串"绿" Play in the "Green" Corridor

泉州养老院设计 3

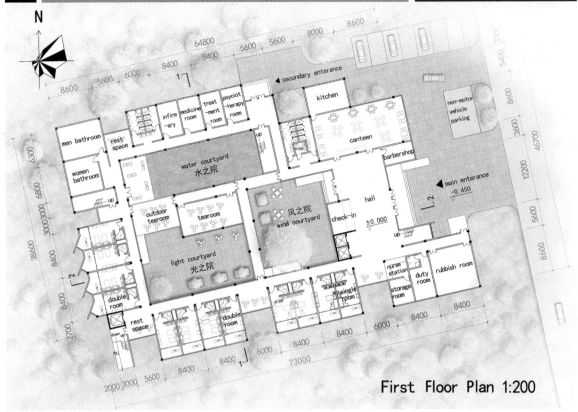

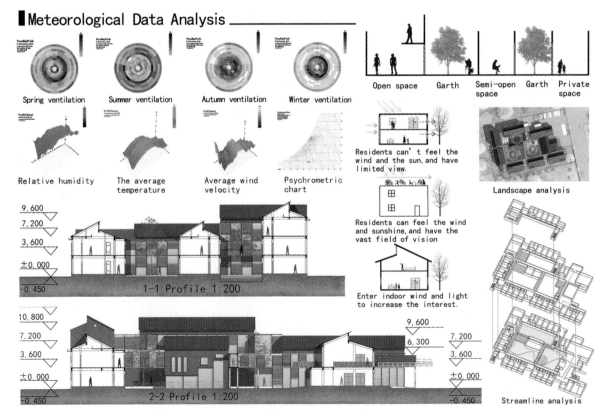

游廊串"绿" Play in the "Green" Corridor   泉州养老院设计 4

Light Courtyard：光影过隙

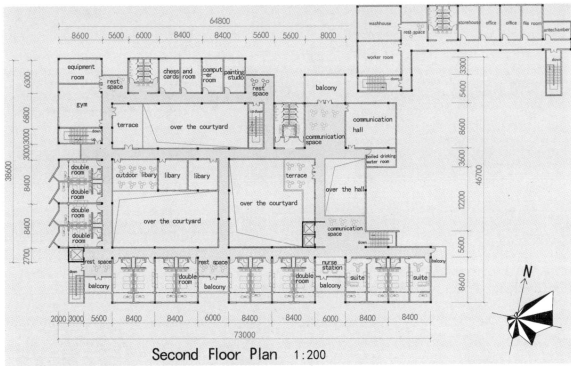

Second Floor Plan 1:200

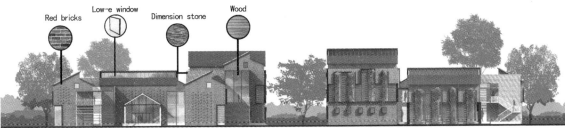

East Elevation 1:200   West Elevation 1:200

游廊串"绿" Play in the "Green" Corridor — 泉州养老院设计 5

Light Courtyard：清风树影

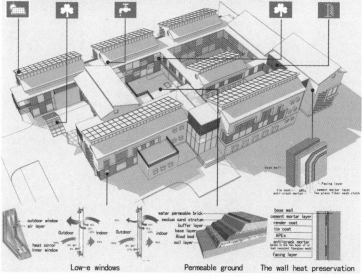

Low-e windows　Permeable ground　The wall heat preservation

The best direction

Solar radiation

Anaerobic Digestion

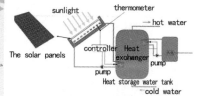

Anaerobic digestion principle not only play the role of dealing with rubbish. The priciple is through the bacteri-al decomposition of organic matter into sugar, then transform into all kinds of acid. Material decomposition after fermentation and produce gas. Bacterial action will generate heat and provid quantitfy heat.

Photovoltaic Power Generation

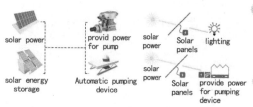

In Quanzhou the average solar radiation amount is 5555.34mwh/m² (Table look-up). Building area 3278.5m². Power consumption index50w/m². Slope roof angle 25°. Annual electricity consumption=50(electricity standard)×3278.5/1000×0.7×8(efficiency)=335062.7kwh/year. Theoreticl annual generating capacity=5555.34×1500(solar panel area)×17.5% (photoelectric conversion rate)=1458276.75MJ=(1458276.75×0.28)kwh =408317.5kwh=40.83million degrees. Due to the actual operation of all aspects of loss. Actual generation is theoretical generating capacity multiplied by actual power generation efficiency=40.83× 0.95×0.89×0.93×0.95×0.88=40.83×65.7%=26.83million degrees.

Water Wall

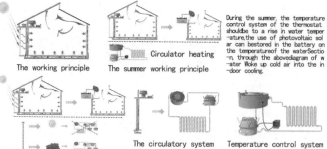

The working principle　　The summer working principle

During the summer, the temperature control system of the thermostat shouldbe to a rise in water temper-ature, the use of photovoltaic solar can bestored in the battery on the temperatureof the waterSectio-n, through the abovediagram of water Woke up cold air into the in-door cooling.

The circulatory system　Temperature control system

In winter, the use of solar radiation heating of water wall, Travel through the circulatory system, the heat cir-culation to the indoor fourWeeks, if meet cloudy day without the sun, te-mperature control The system can usethe solar energysavings of electr-icity or cityPower grid to heat wa-ter.

The winter working principle

Drainage heat recovery system

游廊串"绿" Play in the "Green" Corridor 泉州养老院设计 6

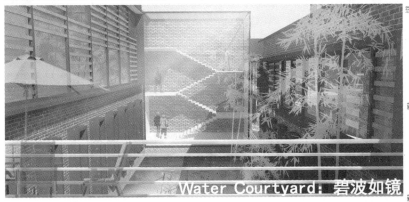
Water Courtyard：碧波如镜

 Frist floor winter 1.5m high wind environment
 Frist floor summer 1.5m high wind environment
 second floor winter 1.5m high wind environment
 second floor summer 1.5m high wind environment
 Third floor winter 1.5m high wind environment
 Third floor summer 1.5m high wind environment
 The overall architecture winter high wind environment
 The overall architecture summer high wind environment

■ Sun-shading Measures

 Vertical greening in combination with the shutters are covered Yang, as shown in the blinds according to The Angle of the sun to adjust, vertical Green is will reuse after rainwater collection Let water flow to drip water its own gravity For irrigation in the tube

- purline 70*70
- drop of water pipe d=40
- window frame
- plant box
- organic soil layer 100 thick
- stomatal slot 25*80
- resistant plants
- outside the shaft d=100
- outside the shaft d=70
- the blinds slice

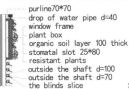 Summer heat   Winter day

■ Ventilation Design

Indoor afforest / Pipe laying

Use of indoor greening to adjust indoor micro climate, make indoor air to purify.

 Indoor selected plants Chlorophytum

Make full use of the courtyard wind and stack effect to improve the wind environment of buildings

The first floor light

The second floor light

Design using the skylight daylighting side window

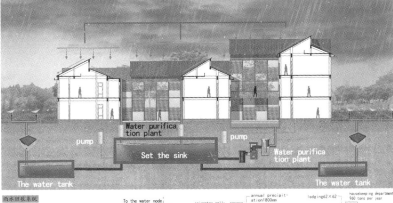
pump — Water purification plant — Set the sink — pump — Water purification plant — The water tank

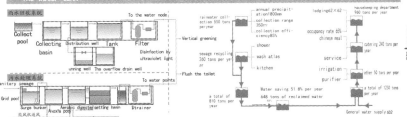
冷水回收系统 / 污水处理系统
Collect pool — Collecting basin — Distribution well — Tank — Filter — To the water node
Disinfection by ultraviolet light
running well / The overflow drain well
Sanitary sewage — Grid pool — Surge bunker — Anoxia pool — Aerobic digester — settling basin — Strainer — To water points

rainwater collection 550 tons per year
collection range 350m
collection efficiency 85%
Vertical greening
sewage recycling 260 tons per year
Flush the toilet

lodging 62×62
occupancy rate 65%
shower / chinese meal / service / irrigation / other 50 tons per year / purifier
catering 240 tons per year
housekeeping department 960 tons per year

a total of 810 tons per year
Water saving 51.8% per year
648 tons of reclaimed water
a total of 1250 tons per year
General water supply 602

太阳能制冷工作原理图
Solar cooling working principle diagram

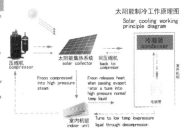
压缩机 compressor / 太阳能集热系统 solar collector / 冷凝器 condenser / 室外机组 / 毛细管 / 室内机组 indoor unit
Freon compressed into high pressure steam
Freon releases heat when passing evaporator turns into high pressure normal temp liquid
Turns to low temp low pressure liquid through decompression
back to compressor

面朝阳光,随"季"应变 Integration Adjustable Bipolar Surface I

综合奖·优秀奖
General Prize Awarded · Honorable Mention Prize

注册号：5292
项目名称：面朝阳光,随"季"应变（泉州）
　　　　　Integration Adjustable Bipolar Surface（Quanzhou）
作　者：王开蕊、苟 堂、陈 希、王丛越、吉俊杰
参赛单位：重庆大学
指导老师：周铁军、宗德新、杜晓宇、张海滨

一、泉州市气候分析　Climate Analysis

太阳能辐射分析　Solar radiation analysis

Analysis:
The best direction is 175 degree,in August,the solar radiant heat is best.

Conclusion:
Building is towards the south by east,we use a justable face in south to shut sun in summer and absort sun in winter.

全年主要气候 Annual climate　　当地朝向分析 Local direction

风向风频分析图　Wind direction frequency analysis

Analysis:
Wind scale is mainly from 2 to 4,wind frequency is mostly southeast and northwest,frequency is changing in summer and stable in transition season.

Conclusion:
Wind energy is well,but wind frenquency isn't stable in summer.We use adjustable face and revolvable shutter to trace the wind.

湿度策略分析图　Humidity Strategy Analysis

Analysis:
In summer,the air is highly hot and wet,natural ventilation isn't enough.

Conclusion:
The building use passitive and mechanical ventilation to improve the comfort level.

被动式湿度调节图　主动式湿度调节图
Passive humidity adjust　Active humidity adjust

舒适度被动设计　Comfort Passive Design

Analysis:
Passive solar heating,high heat capacity material,night ventilation and natural ventilation are profitable for comfort.

Conclusion:
Passive design has an impressive effect in our site.Passive solar heating,heat capacity material,night ventilation and natural ventilation are usable.

Effect to comfort by　Effect to comfort by
pure passive strategy　comprehensive strategy

日光轨迹分析图　Sun-Path Diagram

Analysis:
The main illumination is from the south,solar altitude is pretty hight.It's good for the usage of solar energy.

Conclusion:
The solar energy is rich,and according to the analysis,the location and angle of solar collector can be sure.

综合气候分析　Comprehensive Climate Analysis

Analysis:
Temperature is above comfort level in summer,and below in winter.Humidity is below the comfort level all year.Solar radiant heat is pretty high,especially in summer.

Conclusion:
Reduce solar radiant heat and enhance ventilation in summer.And make measures inversely in winter.

气候指数与人体舒适度对比
Comparison of climate dex and bado comfort level

全年主要气候指数统计

二、区位分析　Location Analysis

方案生成分析　Design process

基地概况分析　Site general situation

基地轴线关系分析　Axis relationship

总平面图（Site-plan）1:750

office entrance　后勤·办公入口
kitchen entrance　厨房入口
建筑主入口　main entrance
site main entrance　基地主入口
室外活动入口　outdoor activities entrance

Technical-economic indicator
技术经济指标：
- Site area　用地面积：4000㎡
- building area　建筑面积：3200㎡
- Floor area　占地面积：1500㎡
- Building density　建筑密度：37.5%
- Floor area ratio　容积率：0.8
- Greening rate　绿地率：40%
- Parking space　停车位：6个

建筑功能区分析　Functional space anlysis

基地太阳能热量分析　Solar radiant heat

Old-age Building Design in Quanzhou

面朝阳光，随"季"应变 Integration Adjustable Bipolar Face II

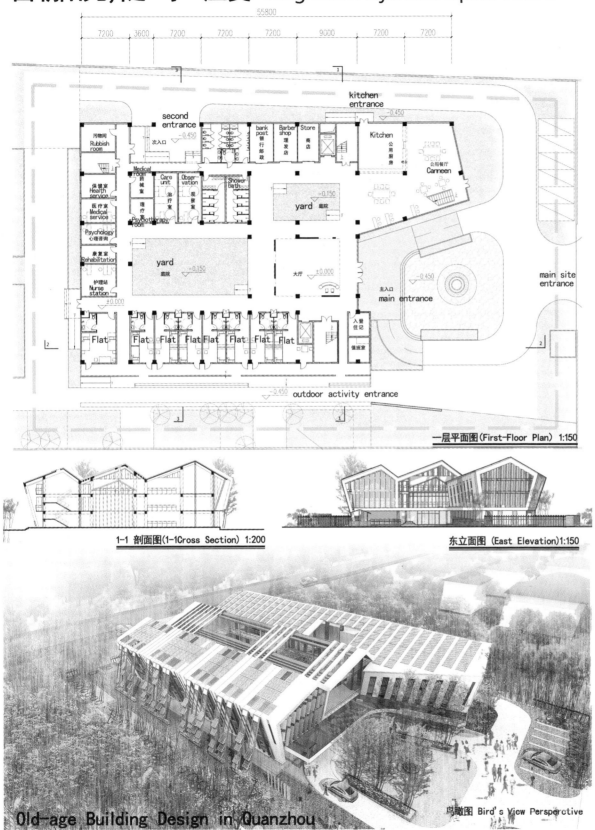

一层平面图 (First-Floor Plan) 1:150

1-1 剖面图 (1-1 Cross Section) 1:200

东立面图 (East Elevation) 1:150

鸟瞰图 Bird's View Persperctive

Old-age Building Design in Quanzhou

面朝阳光，随"季"应变 Integration Adjustable Bipolar Surface IV

主入口人视图 Main entrance perspective

通风策略分析 Ventilate Strategy Analysis

1. 热压通风分析 Ventilation of thermal pressure

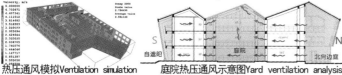

热压通风模拟 Ventilation simulation | 庭院热压通风示意图 Yard ventilation analysis | 庭院热压通风模拟 Yard ventilation simulation

2. 楼梯间拔风通风分析 Wind Raising of Staircase Analysis

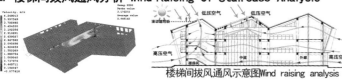

楼梯间拔风通风模拟 Wind raising simulation

楼梯间拔风通风示意图 Wind raising analysis

3. 居住单元通风分析 Ventilation Analysis of Living Apartment

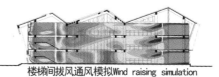

平面风环境模拟 Wind enviroment simulation

房间风环境模拟 Wind enviroment simulation

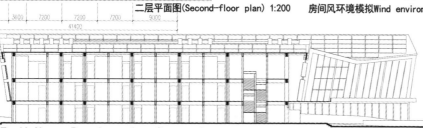

二层平面图 (Second-floor plan) 1:200

2-2 剖面图 (Cross section) 1:200

Old-age Building Design in Quanzhou

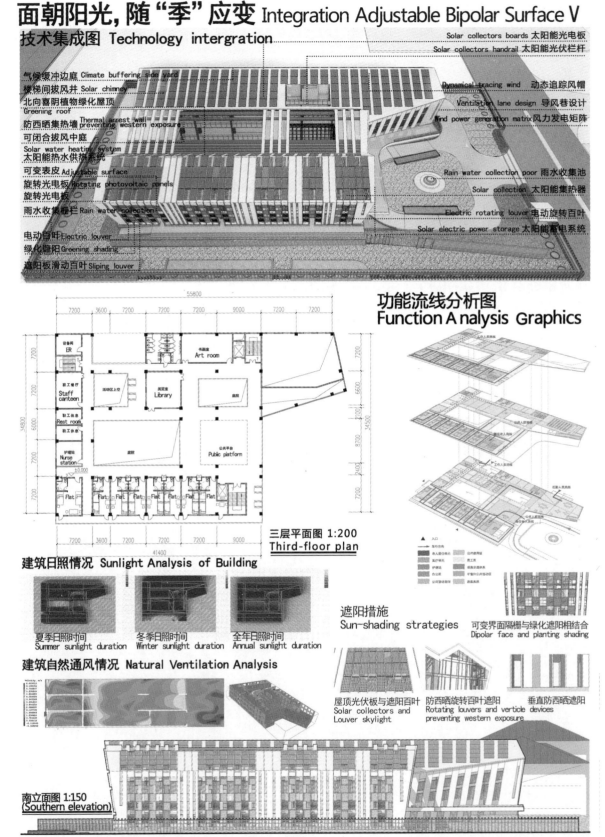

面朝阳光，随"季"应变 Integration Adjustable Bipolar Surface V

Old-age Building Design in Quanzhou

面朝阳光，随"季"应变 Integration Adjustable Bipolar Surface VI

中庭透视图 Perspective of the yard

太阳能水电供需分析
Supply and demand relationship of solarwater and electricity

1.

	用电需求量(Wh)	所需电费(元)	发电量(kw*h)	发电电费(元)
1月	4800	4800	6583.659	3950.19
2月	4800	4800	5840.423	3504.255
3月	4200	4200	6917.619	4150.575
4月	4200	4200	7735.290	4641.18
5月	4200	4200	8518.115	5110.875
6月	4800	4800	8445.169	5067.105
7月	4800	4800	8878.431	5327.055
8月	4200	4200	8672.886	5203.725
9月	4200	4200	7948.320	4768.995
10月	4200	4200	8500.824	5100.495
11月	4800	4800	4869.590	4382.625
12月	4800	4800	6866.018	4119.615
总计	54000	54000	89776.343	55326.69

太阳能光伏发电供需计算 Supply and demand relationship of solarwater and electricity

2.回本年限计算 Payback period
(1). There are 300 solar collectors in total, the photovoltaic power collector costs 20yuan/piece, it will cost about 8000yuan plus installation charge.
(2). When photovoltatic power collectors join the power grid, there wil be generate about 1000yuan every year. And afert 8 years later, the costs will be back, then there will be about 1000yuan of income every year.

太阳能热水供需计算 Supply and demand relationship of solarwater
1. Hot water supply volume
Apartment and public shower cubicle water supply volume: 50*60+8*60*10=7800L
2. Hot water heating supply volume in December-March: about 25000L
3. Solar water cooling supply volume in June-August: alout 25000L
4. Water supply volume in total: about 33000L
Into the Formula:

$$A_c = \frac{Q_w c \rho_r (t_{end}-t_i)}{J_t \eta_{cd}(1-\eta_l)}$$

Conclusion: The need of solar collectors is about 320m², And the true square is about 360m²

Old-age Building Design in Quanzhou

太阳能利用策略分析 Analysis of Energy

太阳能发电情况 Solar electricity generation situation
Solar Silicon wafer photovataic power generation boards are selected, north of abat-vents are filled. Using 1500m*1800m-type boards, adding up to pieces of 149 boards and 400 squares of area.

太阳能利用数据计算 Data of solar utilization
1. Sunshine duration of QuanZhou annual average is 1892~2131 hours. Then the daily average sunshine duration is 5.5 hours.
2. The area of solar power generation boards in this design is about 400 squares, the generated output of every Solar Silicon wafer photovataic power generation board is 3.5W~4W, and the total generated output is Pz=400/0.02436*3.5=57527W. The area of every piece of solar power generation board is 0.02436m², the average daily generating production is W=Pz*T=57527*5.5=316.4Kw*h, then the annual average generating production is 115486Kw*h. Besides the costs of this building and people in here, there are extra productions and earnings can be exist for the extra avtivities.

太阳能技术集成图 Solar technology integration analysis

江水源热泵辅助热源 River heat pump auxiliary thermal source

太阳能热水情况 Situation of solar hot water
Using plate collectors to achieve solar-powered building integration, through the plate collectors in south roof and auxiliary thermal source, the building can product hot water to meet the showering need of the old in summer. During winter, floor heating can operate by the support of solar water heating to warm the living apartment. In summer, solar water can cool the air by involving the air-condition system, as a active strategy, hot water air-condition systerm is a greener way.

自然能的主动策略 Active strategy of natural source

综合奖·优秀奖
General Prize Awarded · Honorable Mention Prize

注 册 号：5461
项目名称：光之寓（泉州）
　　　　　Sanatorium of Sunshine
　　　　　(Quanzhou)
作　　者：郭嘉钰、刘家韦华、郭　璇、
　　　　　时　雯
参赛单位：河北工程大学
指导老师：田　芳、霍玉佼

光之寓 SANATORIUM OF SUNSHINE 1
——泉州生态颐养服务中心设计

设计说明

光之寓，为银龄人提供暖如阳光的居所。方案沿袭我国传统民居形式，以构筑内庭院为空间核心，用楼层形式构建护理单元。合理运用主被动设计策略，运用光伏、光电、光热等多种光能源技术。充分考虑老年人需求，从风、光、热、湿等物理角度和节地节能节水节材等绿色建筑角度出发，营造阳光·颐养的休闲养生环境。

Of sunshine, for the silver age to provide warm as sunshine. The program follows the traditional form of residential buildings in order to build the inner courtyard as the core of space, with the form of building a nursing unit. Rational use of active design and passive design strategy, the use of photovoltaic, photovoltaic, light and other light energy technology, give full consideration to the needs of the elderly, from the wind, light, heat, wet environment and other physical point of view and energy-saving sections of water-saving materials and other green building technology point of view, to create the sunshine·enjoying the leisure and health environment.

Site Analysis
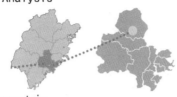

Psychrometric

The effective passive design method is to enhance the regenerative capacity of the structure, night ventilation, natural ventilation, indirect evaporative cooling.

Solar Radiation
   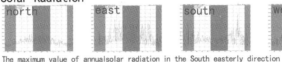

The maximum value of annual solar radiation in the South easterly direction.

Annual Humidity

Most of the time is in a high humidity environment.

Wind Analysis

spring　　winter　　autumn　　summer

The dominant wind direction in summer in this area is southwest wind.

Best Orientation

Best building direction is 182 degrees.

Comprehensive Evaluation
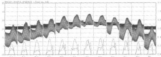

The direct solar radiation is higher in summer and autumn. The annual temperature change is mild.

Conclusions and Strategies

The best building direction is South or South East; Attention should be paid to the natural ventilation organization in the design; the structure design should focus on the wetproof layer design.

Solar radiation resources are rich, photoelectric, light guide and other measures can be used. The precipitation is abundant; so the problem of rain water collection is considered; select appropriate shading facilities.

Formation Analysis

Possibly
Traditional Architecture
Cut
Change
Cumer
Sanatorium of Sunshine

SANATORIUM OF SUNSHINE 2
光之窝 ——泉州生态颐养服务中心设计

2017 台达杯国际太阳能建筑设计竞赛获奖作品集

TECHNICAL-ECONOMIC INDICES
Urban Planning Area: 4927m
Overall Floorage: 3360m
Green Rate: 38%
Building Density: 24%
Volume Rate: 0.68
Bed: 48

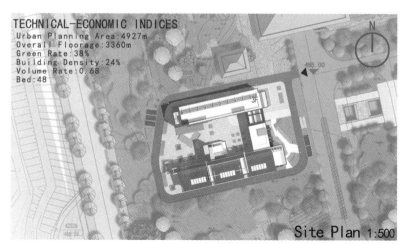

Site Plan 1:500

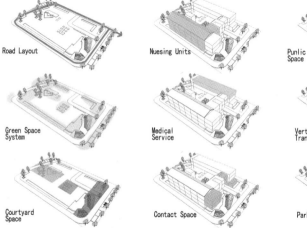
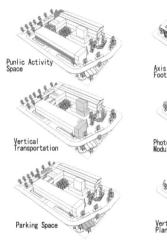
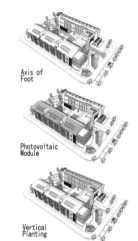

Road Layout — Nuesing Units — Punlic Activity Space — Axis of Foot
Green Space System — Medical Service — Vertical Transportation — Photovoltaic Module
Courtyard Space — Contact Space — Parking Space — Vertical Planting

SANATORIUM OF SUNSHINE 3
光之窝——泉州生态颐养服务中心设计

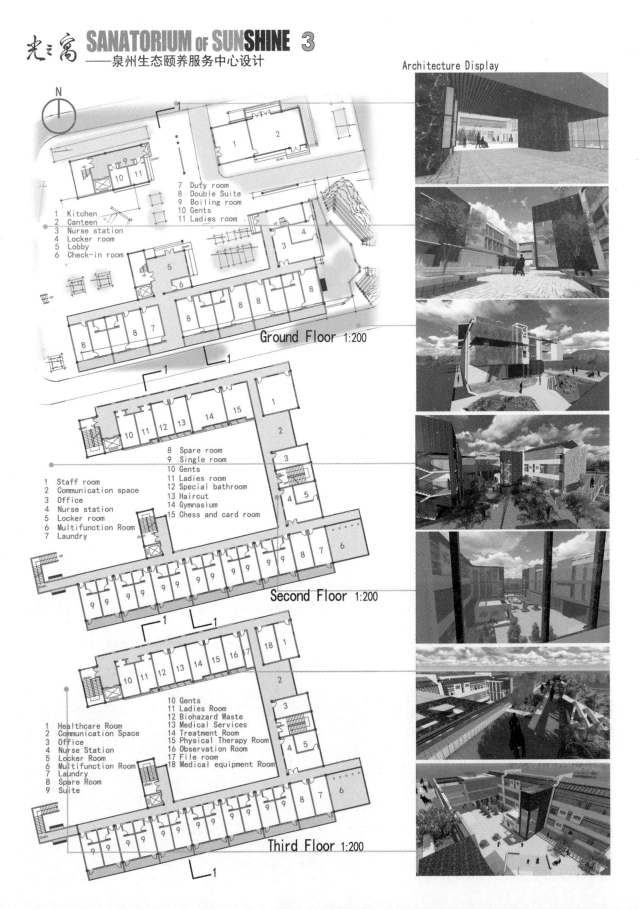

Ground Floor 1:200

1. Kitchen
2. Canteen
3. Nurse station
4. Locker room
5. Lobby
6. Check-in room
7. Duty room
8. Double Suite
9. Boiling room
10. Gents
11. Ladies room

Second Floor 1:200

1. Staff room
2. Communication space
3. Office
4. Nurse station
5. Locker room
6. Multifunction Room
7. Laundry
8. Spare room
9. Single room
10. Gents
11. Ladies room
12. Special bathroom
13. Haircut
14. Gymnasium
15. Chess and card room

Third Floor 1:200

1. Healthcare Room
2. Communication Space
3. Office
4. Nurse Station
5. Locker Room
6. Multifunction Room
7. Laundry
8. Spare Room
9. Suite
10. Gents
11. Ladies Room
12. Biohazard Waste
13. Medical Services
14. Treatment Room
15. Physical Therapy Room
16. Observation Room
17. File room
18. Medical equipment Room

Architecture Display

SANATORIUM OF SUNSHINE 4
泉州生态颐养服务中心设计

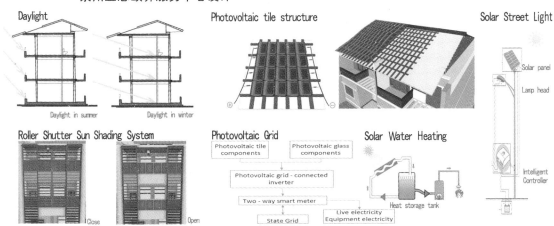

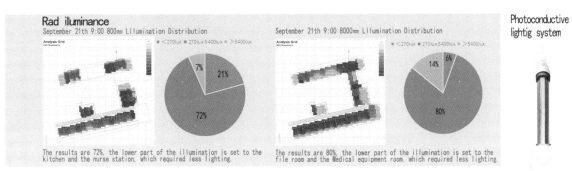

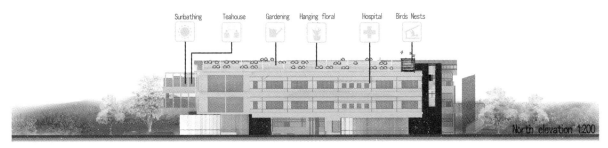

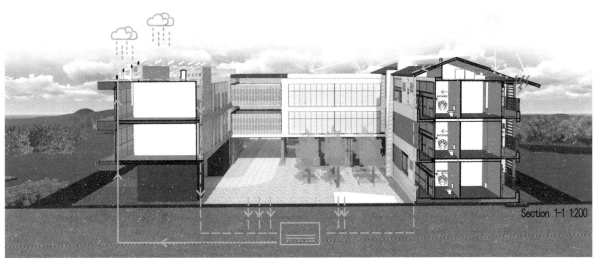

光之窝 SANATORIUM of SUNSHINE 5
——泉州生态颐养服务中心设计

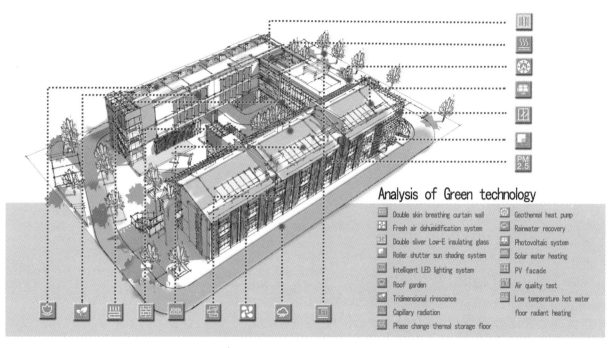

Analysis of Green technology

- Double skin breathing curtain wall
- Fresh air dehumidification system
- Double sliver Low-E insulating glass
- Roller shutter sun shading system
- Intelligent LED lighting system
- Roof garden
- Tridimensional rirescence
- Capillary radiation
- Phase change thermal storage floor
- Geothermal heat pump
- Rainwater recovery
- Photovoltaic system
- Solar water heating
- PV facade
- Air quality test
- Low temperature hot water floor radiant heating

Photovoltaic System
Annual Output of PV Modules Calculation

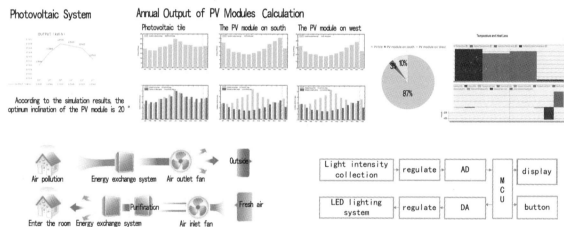

According to the simulation results, the optimum inclination of the PV module is 20°.

Designbuilder Analysis

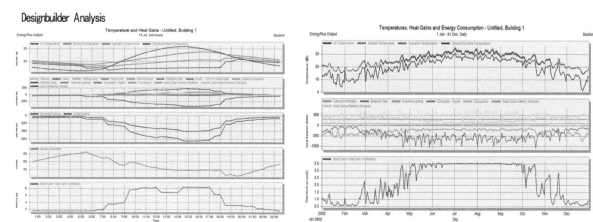

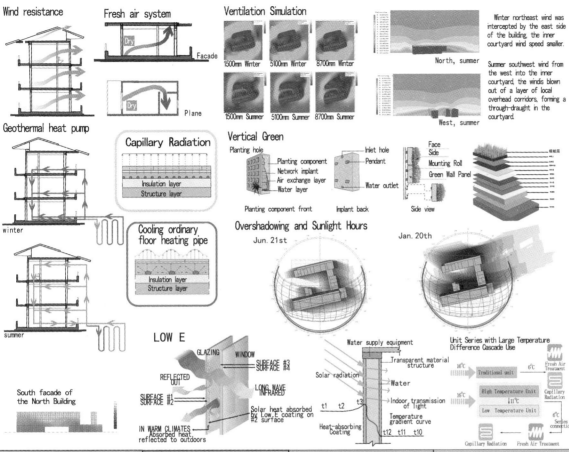

综合奖·优秀奖
General Prize Awarded · Honorable Mention Prize

注 册 号：5513
项目名称：园宅院——伍有宅（泉州）
　　　　　Garden, House, Yard-Five (Quanzhou)
作　　者：张 立、高梦雨、袁 梦、于 露、张 蒙
参赛单位：山东建筑大学
指导老师：张 勤、薛一冰

Bird-view plan

园宅院——伍有园
01 Garden, house, yard-Five

Design Idea (Garden, house, yard-Five)

| minds without words | favorites without scheme | tea without teapot | book without admire | calmness without common |

I was going to explain... but now I forget what it was.

I've got my favorite and more than happy, almost forget the world.

People come, people go.

Through studying success and fortunes will follow. Reading bring us everything.

Without dim not insist your dream, without serenity not go far.

设计说明

建筑的整体布局中设置了多个庭院、天井和街巷，重释也是泉州建筑空间。本案将带有当地特色的井院式布局纳入养老建筑中，选取一组有秩序的院子作为组织空间的主体要素，中心区院子主要承担公共交流，周边院子尺度较小，引入天井概念，用于通风采光。每个院子以自己独特的主题表现，整体呈分散式布置。方案结合当地山地地形，成错落有致之势，利用层与层之间的平台，为老年人留出更多的交往空间。

Design Description

Set up in building the overall layout of multiple courtyards, courtyards and lanes, reinterpretation of quanzhou architectural space. Courtyard style with local characteristics will be the case of Wells to pension building layout, select a set of orderly yard as the main body of organization space elements, central courtyard mainly for public communication, around the yard scale is small, introduced a concept of courtyard, used for ventilation and lighting. Each courtyard with their own unique performance, the theme of the distributed on the whole layout. Scheme combined with the local mountain terrain, into strewn at random have send, use the platform between layer and layer, leaving more communication space for the elderly.

Background Analysis

Site Location Analysis

Leifeng Town Location

In the world of ceramic land - Dehua County of Quanzhou City, Lei Feng Zhen porcelain impression ecological park, north latitude 25 degrees east longitude 118 degrees 56′, 32′, the project is located in central Fujian Province, northwest of Quanzhou.

Quanzhou City　Dehua County　Leifeng Town

The Site　　　The Site　　　The Site

City satellite map　County satellite map　Town satellite map

(1) project is located east and Yongtai County, adjacent to Xianyou County, Putian City, bordering the South and Yongchun County, west of Datian County, North adjoin Youxi County, 7 kilometers from the county seat.
(2) mountains, rich in resources, with "mountain, water, mineral rich, porcelain beauty" four advantages, known as the "Fujian Treasure house".

Site Location Analysis

Dehua county is located in central Fujian roof in Daiyun mountain area, the higher terrain, complex terrain, mountains, mountain to the landscape.

Site Location Analysis

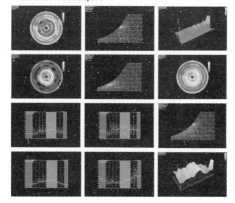

City Image

Urban texture　Quanzhou port　Green building　National culture

园宅院——伍有园
02 Garden, house, yard-Five

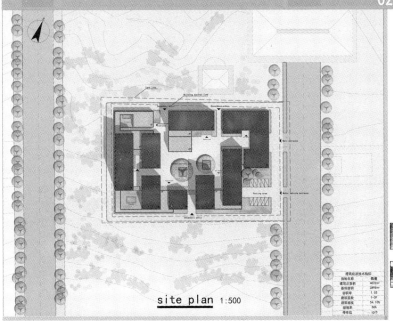

Climatic Analysis

All passive technology | Passive night cooling

Passive ventilation | passive solar heating

Space Construction

Courtship | Conversation | play | Athletic sports | group activity

Quiet | Dancing | Nostalgia | Make friends | Leisure activities

Garden Space Intention

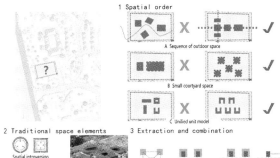

1 Spatial order
A Sequence of outdoor space
B Small courtyard space
C Unified unit model

2 Traditional space elements
- Spatial introversion
- The centrality of space

3 Extraction and combination

Spatial Manipulation

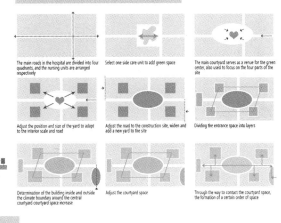

The main roads in the hospital are divided into four quadrants, and the nursing units are arranged respectively

Select one side care unit to add green space

The main courtyard serves as a venue for the green center, also used to focus on the four parts of the site

Adjust the position and size of the yard to adapt to the interior scale and road

Adjust the road to the construction site, widen and add a new yard to the site

Dividing the entrance space into layers

Determination of the building inside and outside the climate boundary around the central courtyard courtyard space increase

Adjust the courtyard space

Through the way to contact the courtyard space, the formation of a certain order of space

Monomer Combination、Courtyard Generation

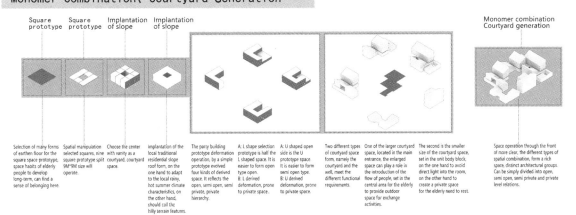

Square prototype | Square prototype | Implantation of slope | Implantation of slope | | | | Monomer combination Courtyard generation

Selection of many forms of earthen floor for the square space prototype, space habits of elderly people to develop long-term, can find a sense of belonging here.

Spatial manipulation selected squares, nine square prototype split 9M*9M size will operate.

Choose the center with vanity as a courtyard, courtyard space.

Implantation of the party building prototype deformation operation, by a simple prototype evolved four kinds of derived space. It reflects the local rainy, hot summer climate characteristics, on the other hand, should call the hilly terrain features.

The party building prototype deformation operation, by a simple prototype evolved four kinds of derived space, semi open, semi private, private hierarchy.

A: L shape selection prototype is half the L shaped space. It is easier to form open type open.
B: L derived deformation, prone to private space.

A: U shaped open side is the U prototype space. It is easier to form semi open type.
B: U derived deformation, prone to private space.

Two different types of courtyard space form, namely the courtyard and the well, meet the different functional requirements.

One of the larger courtyard space, located in the main entrance, the enlarged space can play a role in the introduction of the flow of people, set in the central area for the elderly to provide outdoor space for exchange activities.

The second is the smaller size of the courtyard space, set in the unit body block, on the one hand to avoid direct light into the room, on the other hand to create a private space for the elderly need to rest.

Space operation through the front of more clear, the different types of spatial combination, form a rich space, distinct architectural groups. Can be simply divided into open, semi open, semi private and private level relations.

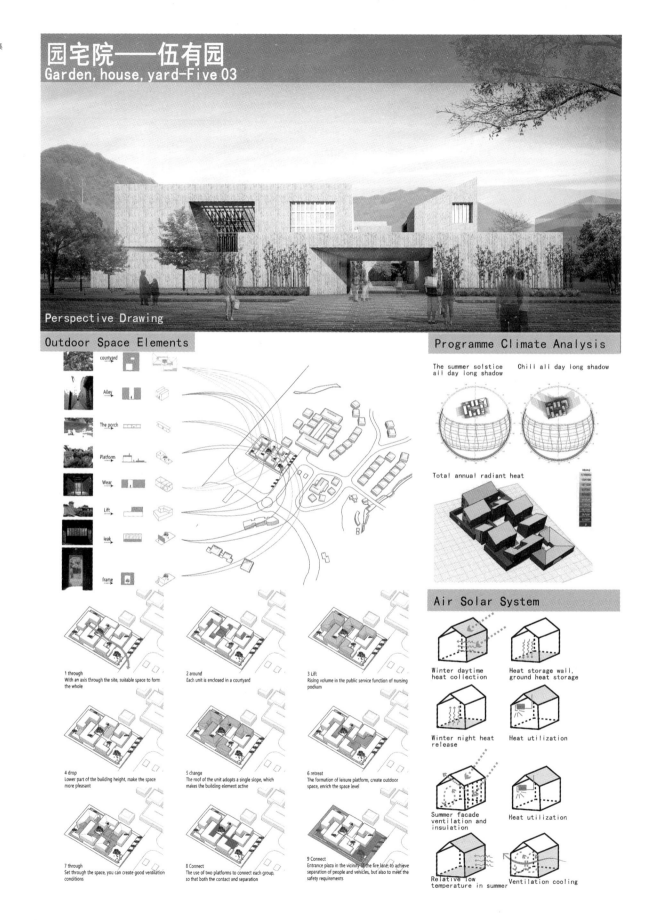

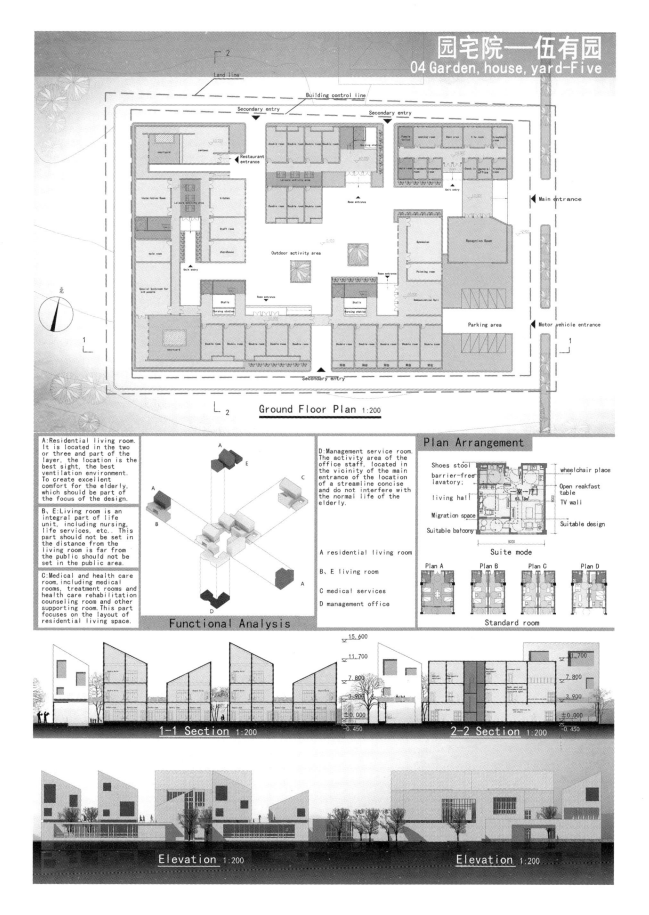

园宅院—伍有园
Garden, house, yard-Five 05

Section analysis

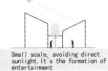

- Small scale, avoiding direct sunlight. It's the formation of entertainment.
- Small gallery can be the cool alley, it's better for cool ventilation.
- The small yard can improve indoor ventilation and lighting and also for playing.
- The roof of the slope can be used to extract the elements of the South Building.
- The corridor between buildings to strengthen the activities of the elderly, reducing the number of stairs up and down.
- The roof height scattered, outdoor platform at different heights, activities for the elderly.
- According to Quanzhou solar elevation angle to design the slope roof, preventing the photovoltaic panels, maximize the use of solar energy resources.
- By subtraction in building hole, improve the air environment inside the building between the increased levels of space.
- The ground floor to form the construction of gray space for the elderly to provide a place to stay and rest.

Industrial assembly technology

Adopting the method of industrial assembly, the separation technology of equipment pipeline and structure is adopted, which is Si System Technology.

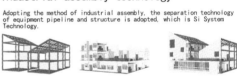

Supporting structure (70-100 years) | Filler block (30-50 years) | Assemble

Prefabricated door, P-toilet module, P-partition, P-floor, P-wall, P-kitchen, P-window, P-railing, P-balcony

Ventilation, sunshade and solar energy integrated vertical sun shading board

Hot season, combined with the principle of hot air, as the power component itself by shading the collected solar energy, driven natural ventilation, which can shade cooling, and promote natural ventilation; both the solar radiation blocking outside the building, and the solar radiation changed for good.

Summer mode

Structure of system — return air grille
Air outlet — Ceramics solar collector
Glass cover — Glass cover
— Indoor air intake
— Outdoor air intake

Second Floor Plan 1:200

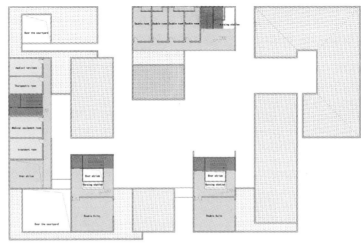

Third Floor Plan 1:200

园宅院—伍有园
06 Garden, house, yard-Five

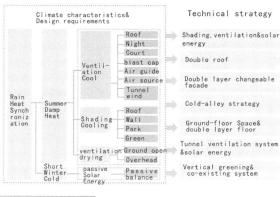

Tunnel Wind Combined Photovoltaic Power Generation System

Tunnel air system: the wet hot air passes through the tunnel to cool, the water vapor condenses out of the air, and the cold air is sent to the room. Photovoltaic power system with wind tunnel system combined with power consumption borne on the wind machine and other mechanical equipment; ventilation and shading plate is heated as the solar chimney, the hot air is dirty from the outdoors, with open wind tunnel design, provide sufficient fresh air into the room.

Construction Methods

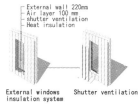

Natural Ventilation

In summer, the main cooling method is ventilation and heat

Quanzhou annual average wind speed of 3.4 m / sec, in autumn (from to December) for the maximum, 4.2 M / s, the southeast monsoon season is weak, the average wind speed is 3.4 m / s.

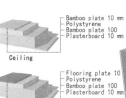

First floor ventilation path map

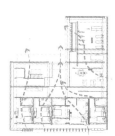

Second floor ventilation path map

Passive Shading and Heating

Balcony can be used to adjust the double deck to achieve different seasons lighting, heating and shading

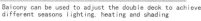

Winter night shutters closed, the heat absorbed during the day to the indoor transmission

A cold winter day, the window down, the formation of a coherent window, lighting and heating

First floor ventilation path map

Summer night, retracted, forming a coherent channel to ventilation and heat dissipation

During the summer, the blinds rotate a certain angle, taking into account the shading and wind guide

First floor ventilation path map

综合奖・优秀奖
General Prize Awarded · Honorable Mention Prize

注 册 号：5524
项目名称：墙之庭院（泉州）
　　　　　The Garden of Walls
　　　　　(Quanzhou)
作　　者：邓佳滢、郑于恬、曾静瑶
参赛单位：广州大学
指导老师：李　丽、刘　源

设计说明：

设计方案以"墙"为灵感，从传统居住群落中学习塑造院落空间的方法并加以创新运用——通过在一片片高低错落、长短不一、不同开洞的墙之间营造庭院与巷道，既满足了绿色建筑的主被动式节能设计的要求，创造了良好、舒适、宜人的生活环境；又让人们能够在绿树繁茂的院落或大或小的空间中穿行相遇、休闲交谈，有利于亲近自然与身心放松。

漫步在光影斑斓的花园中，老人可以在私密、半私密、公共的空间中找到属于自己的那片快乐天地。在这个绿色、低碳、健康养老颐养服务中心，无论是周边福建社区依旧是生态园里的健康或轻度失能的老人，都能在这里找到归属感，感受到家的温暖。

The Garden of Walls
墙之庭院
泉州生态颐养服务中心设计 01

Inspired by the walls : We learn the way to create courtyard space and to use innovatively from the traditional residential communities. Through some pieces of walls with different heights, different lengths, different openings to create some courtyards and roadways, both to meet the main passive energy-saving design requirements of green buildings and create a comfortable, delightful and pleasant living environment at the same time. What's more, people can walk through the lush green courtyard or large or small space and have a leisure conversation, which are beneficial for people to get closer to the nature and their mental relaxation.

Walking in the bright and colorful garden, the elderly can be in private, semi-private, public space to find their own a happy world. Whether it is the surrounding areas of Fujian or ecological park health or mild disability of the elderly, can find a sense of belonging here, feeling the warmth of home in this green, low-carbon, healthy old-age care service center.

MONOMER GENERATE ANALYSIS

1. Live parts is located in the south to accept sufficient sun. Spaces for public activities are in the north side. And to some degree, this will introduce people to get in the venues, so that the elderly are more likely to be encouraged to communicate.

2. Living part was divided into monomers strewn at random with small yard in the front of the monomers. Public courtyard was enclosed by residential area and public area.

3. In concert with the southern design technique, walls with different length are adopted to highlight the yard spaces around the monomer constructions, and at the same time formation three "cold lanes" for natural ventilation.

4. Stilt floor in buildings with roof garden, improve the feeling of people in hot summer yard. There are the holes in the walls which provide fully vision, increase people's contacts between different yards.

5. The adoption of active and passive energy saving technology creates a comfortable indoor and outdoor living environment.

LOOK FORWARD

SOLAR HOUSE ANALYSIS

SUMMER:
With opening the glass folding door in balcony, southeast wind enter indoor refreshingly and visors at the top block most of the sun radiant.

WINTER:
With opening the glass folding door in balcony, cold wind is avoid to blow into the interior. Heat from the sunshine on the glass in balcony is storaged, and radiate to the interior, heating the whole room.

GREEN BUILDING TECHNIQUE ANALYSIS

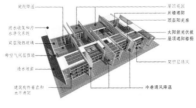

INSIDE THE WALL

create a sense of belonging
Between the two walls is a small living group unit, and surrounded by a small courtyard. The neighborhood of the neighborhood can be familiar with each other; the elderly in the courtyard of the group can do whatever they like.

OUTSIDE THE WALL

shorten the distance between people
By opening the different holes on the walls to increase the opportunity of the elderly in different yards to meet and interact with each other.

From south to north of the ground are private space, semi-private space, public activity courtyard, and public housing.

Publicity increases in turn. The courtyard has shaped the semi-private space perfectly by creating a blurred area between public and private, increasing the likelihood of human activity.

CONCEPT ANALYSIS

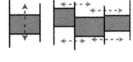

私密空间　　半私密空间　　公共空间

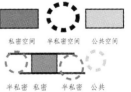

半私密　私密　半私密　公共

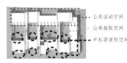

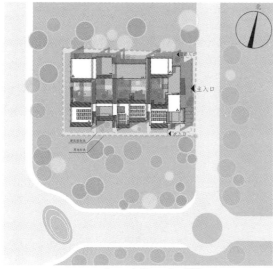

General Plan 1:1000

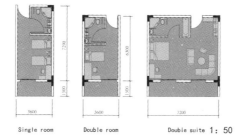

Single room　　Double room　　Double suite 1:50

The Garden of Walls
墙之庭院
02 泉州生态颐养服务中心设计

ECONO-TECHNICAL NORM
Single room:8　　Building density: 40%
Double room:16　　Building area: 3100m2
Double suite:4　　Greening rate:34.2%
Total amount of beds:48　　Floor area Ratio: 0.6

Solar Rediation Analysis

Nouth

East

South

West

First Floor Plan 1:200

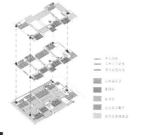

Flow lines and Function Analysis

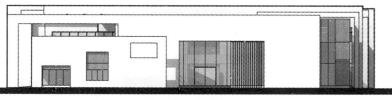

Entrance Elevation 1:200

Southeast Elevation 1:200

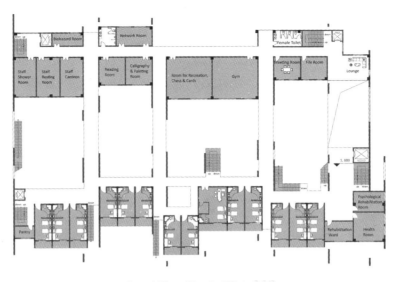

Second Floor Plan 1:200

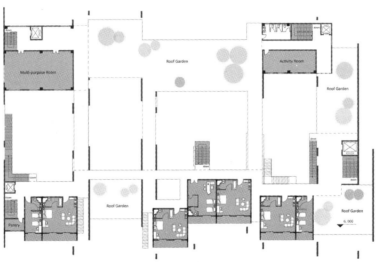

Third Floor Plan 1:200

The Garden of Walls
墙之庭院
03 泉州生态颐养服务中心设计

Hourly Meteorological Data

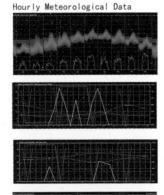

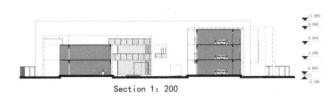

Section 1:200

Solar Radiation Analysis1

The first circumstance
single rooms on 1F and 2F: Horizontal sunshading plus vertical sunshading

The average value of average daily sun radiation: 430.77WH

The second circumstance: solar house on 3F
single rooms on 3F: Horizontal sunshading (glass) plus vertical sunshading

The average value of average daily sun radiation: 478.92WH

single rooms on 3F: Horizontal sunshading (glass) plus shutter sunshading (45 degree level) plus vertical sunshading

The average value of average daily sun radiation: 465.68WH

single rooms on 3F: Horizontal sunshading (LowE glass) plus shutter sunshading (45 degree level) plus vertical sunshading

The average value of average daily sun radiation: 453.37WH

In order to realize the design of solar house to take full advantage of solar energy in winter, the third sunshading plans of solar house with LowE glass has the best effect on reducing sun radiation in summer among the three plans we analyzed.

Analysis of solar rediation in winter

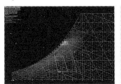
Design one of solar house: Horizontal sunshading (simple glass) plus vertical sunshading
The average value of Average daily sun radiation: 1230.04WH

Design two of solar house: Horizontal sunshading (simple glass) plus shutter sunshading (45 degree level) plus vertical sunshading
The average value of Average daily sun radiation: 1230.04WH1064.61WH

Design three of solar house: Horizontal sunshading (LowE glass) plus shutter sunshading (45 degree level) plus vertical sunshading
The average value of Average daily sun radiation: 1230.04WH1035.41WH

Comparing these three designs of solar house, Design One accepts a maximum value of solar radiation, which makes the best use of solar radiation in winter days with the worst sun shading effect in the summer.
Design Three accept less solar radiation in winter but best sunshade effect in summer. On account of the location in Quanzhou, Design Three is the most appropriate which not only reduce a maximum solar radiation in the summer but also take advantage of solar energy in winter.

Living room daylighting analysis

single rooms on 1F daylighting analysis

single rooms on 2F daylighting analysis

single rooms on 3F daylighting analysis

According to the Daylight factor and Natural Daylighting Illumination calculation: all the rooms meet the requirements of lighting.

The Garden of Walls
墙之庭院
04 泉州生态颐养服务中心设计

Psychrometric Chart Strategy Analysis

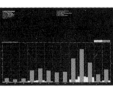

Solar Radiation Analysis2
Sunshade component analysis (courtyard)

Foreword:Predicting "cold lane" and shutter sunshading (45 degree level)will weaken the sun radiation in courtyard. The solar radiation calculation in this architectural design scheme of a courtyard, for example, regardless of the vegetation influence only walls and shutter shade were analyzed.
One:no shutter sunshading and vertical sunshading

The average value of average daily sun radiation: 3251.95WH

Two:vertical sunshading without shutter sunshading

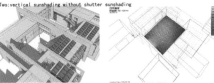

The average value of average daily sun radiation:2464.65WH

Three:shutter sunshading without vertical sunshading

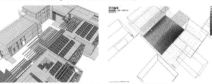

The average value of average daily sun radiation: 2562.67WH

Four:shutter sunshading plus vertical sunshading

The average value of average daily sun radiation: 1824.52WH

Heat radiation calculate for "cold lane"

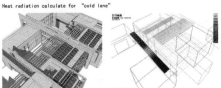

Solar radiation value of cold lane smaller than that in the courtyard, may be able to guide the thermal pressure ventilation.

According to the calculation result, vertical walls made up cold lane effectively reduce courtyard sun radiation, provide a better environment for the elderly to activity in the yard. To combine beauty and technique, we adopt "cold lane" with 45 degrees horizontal shutter sunshading in the courtyard as a sun shading plan.

energy consumption analysis1
material assignment and zone management

1—20mm 混合砂浆；
2—240mm 混凝土多孔砖，界面剂；
3—聚苯颗粒保温浆料；
4—3mm 抗裂砂浆（网格布）；弹性底涂，柔性腻子，外墙涂料

Walls

1—防水层
2—20mm 水泥砂浆找平层
3—最薄 30mm 轻骨料混凝土找坡层
4—100mm 加气混凝土砌块保温层
5—挤塑聚苯板
6—钢筋混凝土面板

Roof

1. Glass Standard
2. Air Gap
3. Glass Standard

Window

material of Retaining structure plan two

1	烧结砖	110.0
2	空气间层	50.0
3	混凝土	110.0
4	找平层	10.0

Walls

2	薄膜	2.0
3	空气间层	150.0
4	薄膜	2.0
5	玻璃纤维	50.0
6	水泥砂浆	10.0

Roof

	标准玻璃	6.0
	空气间层	30.0
	标准玻璃	6.0

Window

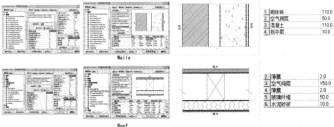

zone management schedule of individuals in room schedule of Lighting and small appliances

The Garden of Walls
墙之庭院
05 泉州生态颐养服务中心设计

energy consumption analysis 2
Analysis of energy consumption:plan one

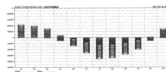

Analysis of energy consumption:plan two

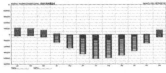

制冷能耗对比   采暖能耗对比

■ COOLING (Wh)材质一 — MONTH ■ COOLING (Wh)材质二   ■ HEATING (Wh)材质一 — MONTH ■ HEATING (Wh)材质二

逐月对比

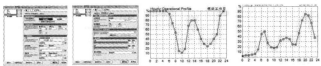

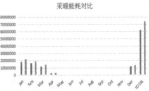

According to the diagram, material plan one making a lower energy consumption of heating than material plan two,but a higher energy consumption of refrigeration. Reason could be that plan one has a higher heat storage capacity,which is warm in winter,but excessively hot in summer, causing a larger energy consumption in total. Therefore the plan two is more appropriate in Quanzhou.

Indoor Wind Environment Analysis

First Floor

Second Floor

Third Floor

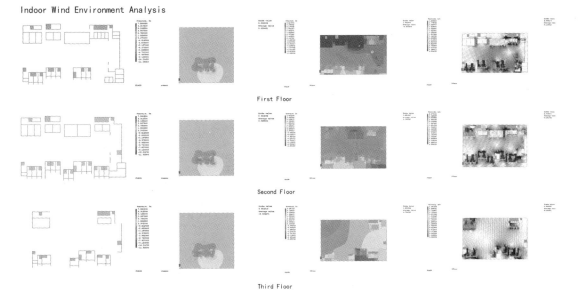

The Garden of Walls
墙之庭院
06 泉州生态颐养服务中心设计

Renewable Energy Analysis
Renewable energy in this program mainly uses solar energy and wind energy.

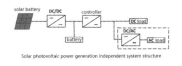
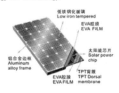
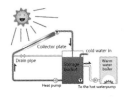
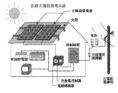

solar energy
(1) Shenzhen annual average solar radiation is 1427.15kWh/m2.
(2) Total building area in this program is 3470m2, according to < Code for planning in city stores GB 50293-1999>, the building electricity consumption target is 65W/m2.
(3) The annual electricity consumption of the building Q=total building area x annual electricity consumption=3470x0.5x8x365x65= 329303kwh/year
(4) Choice of square angle: Consider the geographical location of Shenzhen, at the same time to facilitate the square matrix bracket design and installation convenience, generally take the integer angle, combined with the standard recommended value, and finally get the installation angle of 30 degrees.

(5) The annual electricity generation per square meter of photovoltaic panels=1427.15x0.15x0.75=160.6(kWh/m2·year)
(6) The design of the photovoltaic board area of 450 square meters, so the annual power generation of the building is 160.6(kWh/m2·year)x450m2=72270kWh·year
(7) All solar cells in this design are made up of several pieces of 100WpHYT100D-24 monocrystalline silicon solar standard components, size of the monocrystalline silicon solar standard components is 1000mmx750mm.

Outdoor Wind Environment Analysis

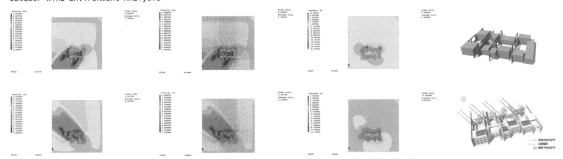

综合奖·优秀奖
General Prize Awarded · Honorable Mention Prize

注 册 号：5584
项目名称：卷光帘（泉州）
　　　　　Sunshine · Production ·
　　　　　Golden-ager（Quanzhou）
作　　者：吴浩然、徐墨林、吴宽、
　　　　　邵丹
参赛单位：天津大学建筑学院、南开大学
　　　　　外国语学院
指导老师：张玉坤、赵劲松

Site Analysis

The place enjoys a subtropical monsoon climate characterized by hot and rainy summers and cold and dry winters with an annual average temperature of 18.2 ℃.

Conclusion & Strategy
1. The building's optimal orientation is 10 degrees east of south.
2. Protections from rainfall and wind, especially southeast wind in summer are to be emphasized and attention is to be paid on adequate natural ventilation in autumn and winter in passive design building.
3. It is recommended that photovoltaic power generation technology should be employed taking into accounts of rich radiation amount.
4. Building components are suggested to be used to collect rainwater.

Climate Analysis

Annual solar radiation and temperature change maps shows that the solar radiation reaches to the highest point at noon in summer, autumn and winter while maintains a low standard in spring. In summers the temperature is comparatively high at noon and moderate at the rest of time.

Southeastern radiation maintains a relatively high amount and remains steady reliable through the whole year, while western radiation has a seasonal fluctuation with higher peaks from the solar radiation map. Therefore more attention is to be paid on protections from western solar radiation.

Living Situation of the Elderly

Peers' Company | Healthy Production Activities | Public Communication Space | Careful Medical Treatment

Design Starting Point

In the face of the high social cost of the old-age care, it is of great urgency to focus on the issue of how to promote the sustainable development of the old-age care and community construction. This project proposes a strategy of productive community for the elderly: by providing the elderly with opportunities to grow vegetables and fruits, this strategy satisfies the elderly's physical and mental needs. The old people can make full use of their rest life. This approach fully mobilizes the retired people's passion to work, and makes community a unit within the production-consumption integration distributed network of resource.

Intention Picture

The psychometric chart and optimal orientation graphic suggest that the optimal orientation of the building is 10 degrees east of south. Natural ventilation fits into the local environment as a passive technology.

From the point of view, the site is rich in wind resources, especially in summer. It prevails sortheast wind in spring and summer, and it prevails northeast wind in autumn and winter.

Site status

卷光帘 | SUNSHINE · PRODUCTION · GOLDEN-AGER 01

Strategy & Characteristics

1. Photovoltaic garden: The combination of transparent photovoltaic and plantation is the subject of the shaping of community landscape. Differ from the old-fashioned landscape greening, the productively continuous landscaping is adopted to develop dots and patches distributed vegetarian parks and fruit gardens.
2. Continuous courtyards: Inner courtyards directly connect to lounge bridges, which maintains the resident's privacy and reflects the domestic environment's endocentric nature.
3. Green Technology: Building components including transparent photovoltaic film, dotted plating brick, automated sun-shading louver, wood planting frame, and photothermal roof are designed specifically to create an ecological and environment-friendly green space for the elderly to live a comfortable and healthy life.

Econo-technical Norms

Total Building Land Area	4928㎡
Total Floor Area	3450㎡
Building Coverage Area	1735㎡
Plot Ratio	0.70
Green Ratio	75%
Building Coverage Ratio	35%
The Number Of Parking Spaces	7

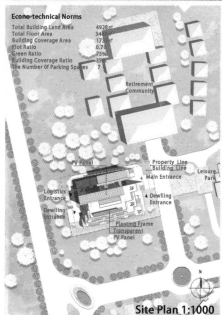

Site Plan 1:1000

Design Specification

本方案将交通、建筑、农业与退台有机整合，通过生产性元素的置入以组织多种公共活动空间布局，打造适应老年人生活习惯的养老建筑。

The concept of design is an organic integration of transportation, architecture, agriculture and backwards terrace. Imbedded with productive elements, public space is organized with a creative diversity. The residential care-home is built to adapt to living habits of the elderly from every aspects.

Building Prototype

Inspired by a Holland architects' future urban model in 1980s, the design has developed an inward, cordial, open, safe and environment-friendly living circumstance and a three-dimensional, open and active public space integrating buildings and transportation, by virtue of drawing on comprehensive advantages of traditional courtyard houses, cave dwellings and terraced planting.

Analysis of the Elderly

- **Motivation Analysis**
  - medical facilities
  - independence
  - quality of life
  - not loneliness
- **Behaviors Analysis** (Conventional)
  - sleeping / chating / reading / TV&radio
  - walking / basking / checking / exercising

Q: restraining potential passive self-help to be easily lonely...
+ communication producting pastoral life...

(innovative)

- **Concept Analysis** — Breaking barriers

Planning Analysis

Traffic analysis | Fire analysis | Function analysis

Formation Analysis

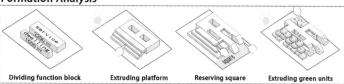

Dividing function block | Extruding platform | Reserving square | Extruding green units

Solar Radiation Analysis

THeat insulation measures are the key point of the design because of intense solar radiation. Meanwhile the design of the three connected courtyard fits into the climate environment of the south. Besides the inner courtyard still receive abundant natural lighting and external reflection.

Total Radiation | Total Direct Radiation

Total Diffuse Radiation | Uniform Sky Factor

Summer Solstice 6.21 | Autumn Solstice 9.21

Winter Solstice 12.21 | Spring Solstice 3.21

Sunshine Range Analysis

The base is located near the tropic of cancer, at the subtropical region which is characteristic of long duration of sunshine. From the sun shadow analysis of solstices arriving time, atriums and rain wind corridors are suggested to establish to bring comfortable experience of space. Meanwhile with an attempt to make a full use of solar energy, corresponding plants are suggested to grow based on the time and place of sunshine.

02 卷光帘 SUNSHINE PRODUCTION GOLDEN-AGER

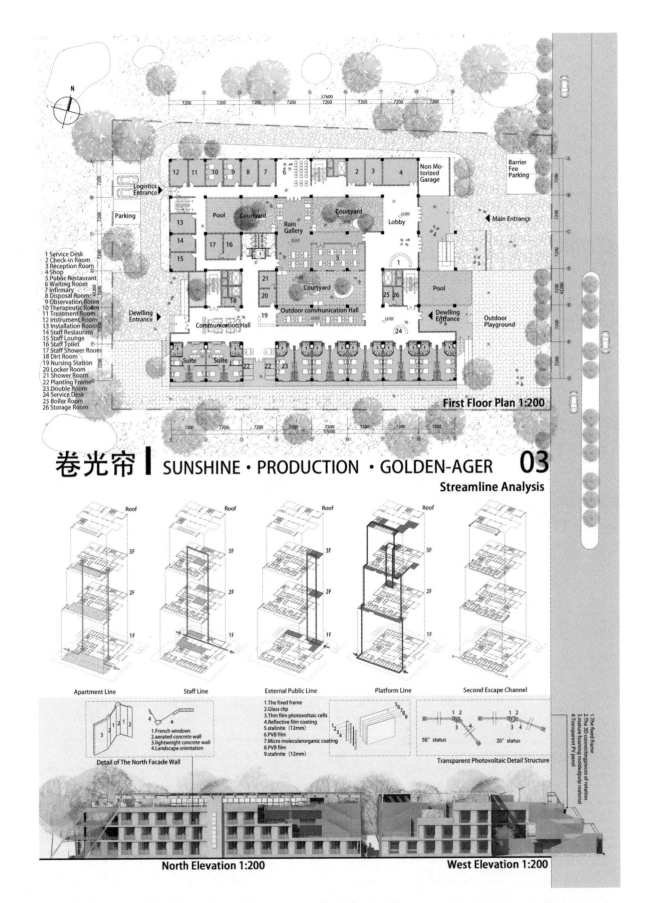

卷光帘 | SUNSHINE · PRODUCTION · GOLDEN-AGER 04

Residential Unit Analysis

1. Unit Style Analysis

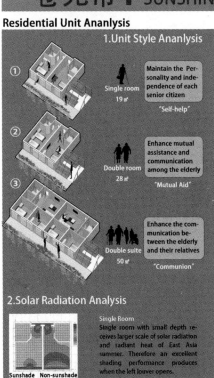

① Single room 19 ㎡ — Maintain the Personality and independence of each senior citizen — "Self-help"

② Double room 28 ㎡ — Enhance mutual assistance and communication among the elderly — "Mutual Aid"

③ Double suite 50 ㎡ — Enhance the communication between the elderly and their relatives — "Communion"

2. Solar Radiation Analysis

Single Room
Single room with small depth receives larger scale of solar radiation and radiant heat of East Asia summer. Therefore an excellent shading performance produces when the left louver opens.

Double Room
Double room, with its large space and depth, receives most of radiant heat on sun-facing side. It gives an outstanding performance as heat insulation measures are applied.

Double Suite
Double suit has a reasonable distribution of space. Each senior citizen owns exclusive rest room. And given sunlight's influence on rest room, Louvers when opened at proper time will help the aged to rest and relax.

Function Volume Analysis

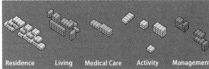

Residence | Living | Medical Care | Activity | Management

Planting Frame Perspective

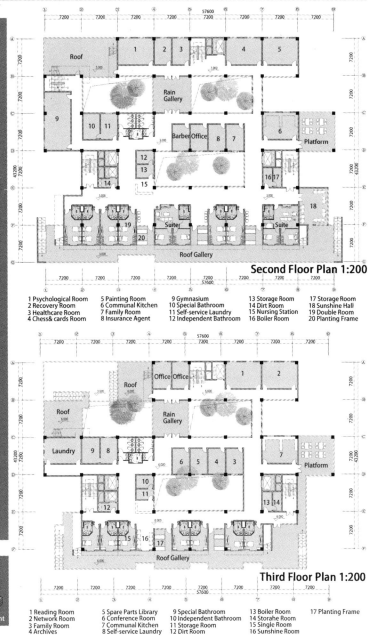

Second Floor Plan 1:200

1 Psychological Room
2 Recovery Room
3 Healthcare Room
4 Chess & cards Room
5 Painting Room
6 Communal Kitchen
7 Family Room
8 Insurance Agent
9 Gymnasium
10 Special Bathroom
11 Self-service Laundry
12 Independent Bathroom
13 Storage Room
14 Dirt Room
15 Nursing Station
16 Boiler Room
17 Storage Room
18 Sunshine Hall
19 Double Room
20 Planting Frame

Third Floor Plan 1:200

1 Reading Room
2 Network Room
3 Family Room
4 Archives
5 Spare Parts Library
6 Conference Room
7 Communal Kitchen
8 Self-service Laundry
9 Special Bathroom
10 Independent Bathroom
11 Storage Room
12 Dirt Room
13 Boiler Room
14 Storahe Room
15 Single Room
16 Sunshine Room
17 Planting Frame

Adaptive Analysis of the Elderly

Family Reunion

Increase Warmth

Increase Security

Observation Line

Promote Communication

Promote Communication

Nursing Line of Sight

Barrier-free

SUNSHINE · PRODUCTION · GOLDEN-AGER 05

卷光帘

Water Circulation System 1:200
— Rain Water   — Irrigation Water
--- Clean Water --- Solar Hot Water

The Effect of Transparent PV
1. Greenhouse
2. Absorb direct sunlight
3. The elderly bedroom
4. The elderly bedroom

(summer)   (winter)

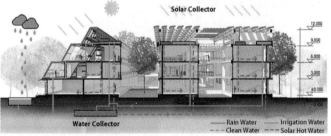

Restaurant
Supermarket
Organic market
Direct selling

Chinese Trumpet Creeper · Sedum Lineare · Berberis Thunbergii · Ligustrum Quihoui · Calathealeopardina · Grape · Dimocarpus longan · Peas · Cherry Tomatoes · Cucumber · Eggplant

Ivy · Erythrina · Boston Ivy

The Elderly's Productivity

Productive building is defined as a multi-level system, which also integrated with the functions of agriculture production, energy production and social & cultural capital preservation and redevelopment. Through this design, the productive strategy is proposed. We hope to blend the energy production, food production and organic elderly community relationship. It promotes the sustainable development of the care center and provide a new possibility.

East Elevation 1:200   **South Elevation 1:200**

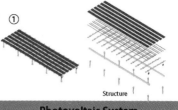
Structure
Photovoltaic System
wind power generation
Rainwater collected by the brick can permeate through the filter screen between the tank and the planting groove.
solar water heating
Other Energy Collection Systems
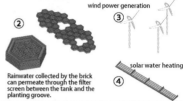
Winter / Summer

On the terrace / On the roof / On the terrace
⑦ Three sizes of planting frames are built for the elderly to walk easily among the plantings.

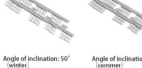
Angle of inclination: 50° (winter) Angle of inclination: 10° (summer)
⑥

⑤ Automated Sunshade Louvre

Planting Frames | **Transparent PV Panel** | **Automated Sunshade Louvre**

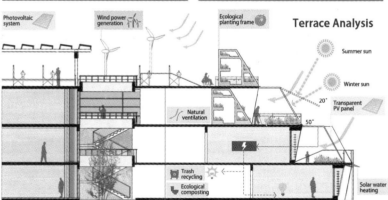
Terrace Analysis
Photovoltaic system / Wind power generation / Ecological planting frame / Summer sun / Winter sun / Transparent PV panel / Natural ventilation / Trash recycling / Ecological composting / Solar water heating

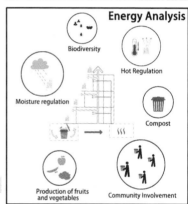
Energy Analysis
Biodiversity, Hot Regulation, Moisture regulation, Compost, Production of fruits and vegetables, Community Involvement

Atrium Perspective

Ventilation Analysis

The south China is hot through the whole year. Given its intense sunlight, the interconnected courtyard is designed to introduce cold air and reduce the exchange of heat to achieve insulation effect.

Rainfall Activities Analysis

By building several rain gallery and grey space, courtyards and corridors enable the elderly to go for a walk outside even in rainy days.

卷光帘 | SUNSHINE · PRODUCTION · GOLDEN-AGER   06

综合奖·优秀奖
General Prize Awarded · Honorable Mention Prize

注　册　号：5587
项目名称：颐养苑——泉州生态颐养服务中心（泉州）
　　　　　The Blissful Pure Land (Quanzhou)
作　　　者：裴婉煜、李靖、胡平、张雅琳、杜衍旭、郭嘉钰
参赛单位：河北工程大学
指导老师：王晓健、霍玉佼

颐养苑 泉州生态颐养服务中心设计 2
The blissful pure land

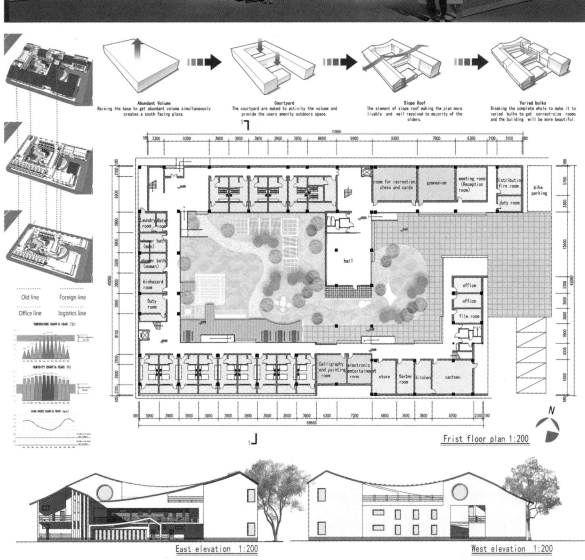

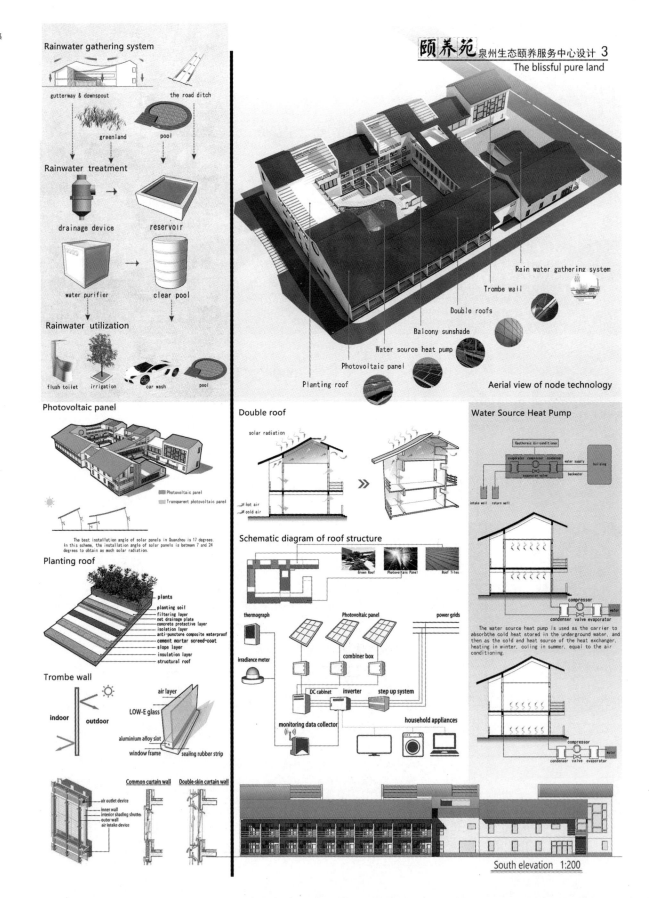

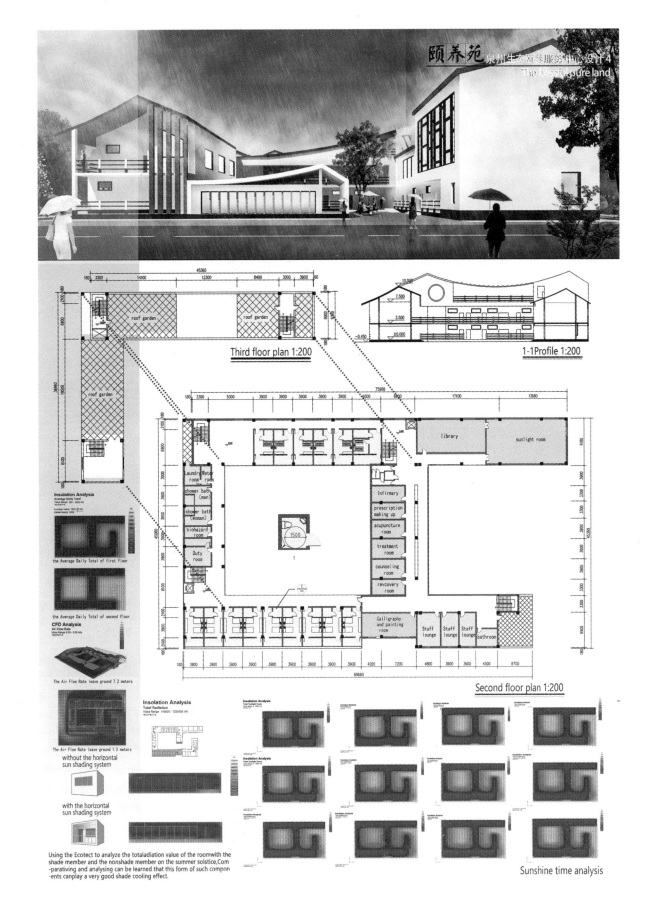

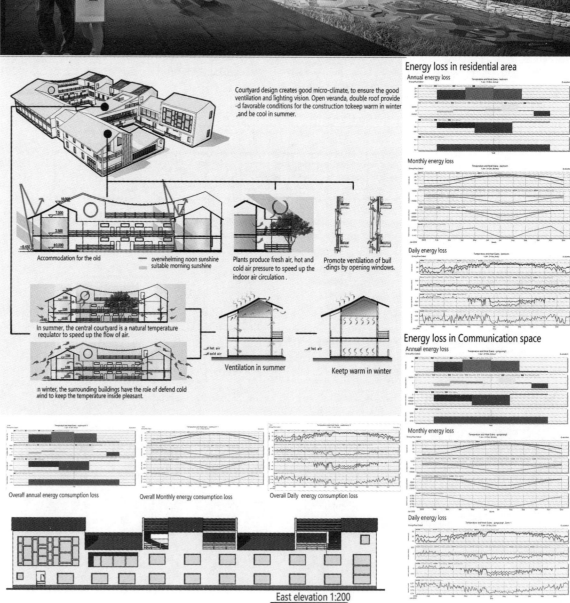

颐养苑 泉州生态颐养服务中心设计 6
The blissful pure land

Green building rating

Land and outdoor environment
Q1 61

Control item		content
Scoring item	Economical and intensive use of land	19
	Rationally set up green land	9
	Avoid light pollution	4
	Environmental noise standard	4
	Good wind environment	4
	Green rainwater infrastructure	9
	Planning surface and roof runoff	6
	Reasonable choice of green way	6

Energy conservation and energy utilization
Q2 65

Control item		content
Scoring item	Building optimization design	6
	The window can be opened to achieve good ventilation effect	6
	The thermal performance of the envelope is better than the national standard	10
	The cold and heat source of air conditioning system is superior to the national standard	6
	Reasonable selection and optimization of heating, ventilation and air conditioning system	7
	Measures to reduce energy consumption in heating, ventilation and air conditioning systems in the transitional season	6
	Orientation, subdivision heating, air conditioning area	6
	Lighting power density value to the current national standard	8
	Rational utilization of renewable energy	10

Water saving and water resources utilization
Q3 75

Control item		content
Scoring item	The average daily water consumption of the building meets the current national standards for water saving	10
	Take effective measures to avoid leakage	7
	The water supply system without overpressure the flow	8
	Water metering device	6
	Sanitary appliances with high water use efficiency	10
	Greening irrigation use water saving irrigation	10
	Air conditioning equipment or system using water cooling technology	10
	Rational use of unconventional water resources	7
	Cooling water replenishment using unconventional water	7

aterial and material utilization
Q5 63

Control item		content
Scoring item	Selection of architectural form	9
	Prefabricated industrialized production	4
	Integrated design of the kitchen, bathroom	6
	Selection of locally produced building materials	8
	Cast-in-place concrete with mixed concrete	10
	Building mortar with mortar	5
	Rational use of high-strength building materials	8
	Rational use of high durability building materials	5
	Use of recyclable materials and recyclable materials	8

Indoor environmental quality
Q5 63

Control item		content
Scoring item	Main function room noise level	6
	Sound insulation performance of the main function room	8
	The main function room lighting coefficient to meet the current n	8
	Improve indoor daylighting effect of building	10
	Adopt adjustable shading measures to reduce the heat of the sun	6
	Heating and air conditioning system at the end of the site can be independently adjusted	8
	Optimization of architectural space, layout and structural design to improve the effectiveness of natural ventilation	10
	Reasonable airflow organization	7

$\Sigma Q = w_1 Q_1 + w_2 Q_2 + w_3 Q_3 + w_4 Q_4 + w_5 Q_5 = 66.21$ (Two stars green building)

The architectural shadow changes at different times in a year

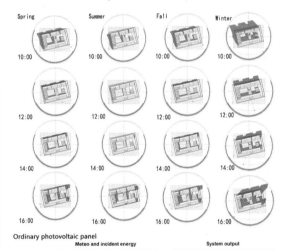

Ordinary photovoltaic panel

	Gl. horiz. kWh/m2.day	Coll. Plane kWh/m2.day	System output kWh/day	System output kWh
Jan.	2.76	3.33	397.2	12314
Feb.	2.72	3.03	361.4	10119
Mar.	2.88	3.02	360.1	11165
Apr.	3.60	3.59	428.7	12861
May	3.87	3.71	442.7	13725
June	4.30	4.03	481.2	14437
July	5.32	4.99	595.4	18457
Aug.	4.67	4.57	545.5	16912
Sep.	3.87	4.01	478.1	14344
Oct.	3.57	3.97	473.5	14678
Nov.	3.07	3.67	437.8	13133
Dec.	2.94	3.70	441.4	13684
Year	3.64	3.81	454.3	165828

Transparent photovoltaic panel

	Gl. horiz. kWh/m2.day	Coll. Plane kWh/m2.day	System output kWh/day	System output kWh
Jan.	2.76	3.33	27.11	840
Feb.	2.72	3.03	24.66	691
Mar.	2.88	3.02	24.58	762
Apr.	3.60	3.59	29.26	878
May	3.87	3.71	30.21	937
June	4.30	4.03	32.84	985
July	5.32	4.99	40.63	1260
Aug.	4.67	4.57	37.23	1154
Sep.	3.87	4.01	32.63	979
Oct.	3.57	3.97	32.31	1002
Nov.	3.07	3.67	29.89	896
Dec.	2.94	3.70	30.12	934
Year	3.64	3.81	31.00	11316

综合奖·优秀奖
General Prize Awarded · Honorable Mention Prize

注 册 号：5607
项目名称：暮邻·乐居（泉州）
　　　　　Live with Happiness in Old Age（Quanzhou）
作　　者：宋智霖、宣勤朗、戴安李、孙力枰、周　立
参赛单位：厦门大学
指导老师：罗　林

暮邻·乐居 ——泉州生态颐养中心
LIVE WITH HAPPINESS IN OLD AGE

设计说明
本案立足于泉州当地气候条件和基地环境，注重太阳能、风能、水能等资源的利用，采用多种主动技术与被动技术改善室内环境，同时结合老年人的生活与活动特点对建筑进行设计，采用蓄热、通风等节能技术，致力于为老人提供一个更加绿色、舒适的环境。设计提取闽南建筑元素，以回游动线的流线型是形成庭院围合空间，为老年人在建筑中游走，停留，交谈等提供了众多方便交流的空间。满足老人心理需求，做到暮邻友好，安居晚年。

Design statement
This case on the local climate conditions and the base environment of Quanzhou, we pay special attention to the use of resources such as solar, wind, and water. In order to give the old a more green and comfortable health environment, we have taken the characteristics of life and activities of the old into consideration. Therefore, we have adopted the energy conservation technology such as heat storage, ventilation and so on. What's more, like Minnan's traditional buildings, we use some courtyards in our design to provide more comfortable communication space for the elderly. We are committed to provide a good environment for them live with a happiness in old age.

Geography Analysis

Climate Analysis
According to climate analysis, the main wind direction in summer is southeast wind, and in winter the frequency wind is northwest wind. And it have haavy air humidity and high temperature during whole year. Theremore, Quanzhou have a perfect conditions of useing solar energy.

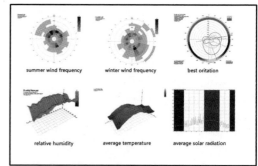

Site Analysis
The site is located in the west of an ecological center in the porcelain city Dehua, where is richly vegetated, and there is beautiful landscape, fresh air and plenty of thermal springs. Simultaneously, there are solid infrastructure and convenient traffic condition, which benefit the construction and the elders' life.

Behavioural Analysis
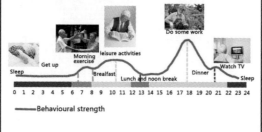

According to the above analysis, the elders spend much time on collective activities such as playing chess and cards and enjoying the sunshine. We concern more about daily communication of the elders, free barrier design as well as their healthcare in our work.

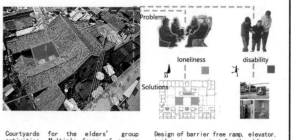

Courtyards for the elders' group activities. Multiple forms of spaces including outdoor space, interior space and the transitional zone, make various activities happen.

Design of barrier free ramp, elevator, toilet, is convenient for the elders to use.

暮邻·乐居 ——泉州生态颐养中心
LIVE WITH HAPPYINESS IN OLD AGE

Block Generation

① A Dacuo consists of a main courtyard and two on both sides. In this design, we remove the living room that is public space and for worshipping.

② without the living room, there are one large middle yard and small ones on either side.

③ copies of the unit. Jutouzhi prioritize spacial patterns with large yards and small ones.

④ yards of clear priorities, offer the elders enough living space to make them enjoy their life.

⑤ the roof incorporate all the units to an integrated whole.

⑥ the hallway and concourse hall divide the yards into two parts

Suite Room Plan 1:100

Double Room Plan 1:100

Shadow Analysis
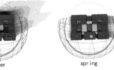
winter — spring

autumn — summer

First Floor Plan 1:200

1. concourse hall
2. hallyway
3. countyard
4. balcony
5. registration room
6. nurse station
7. preparation room
8. laundry room
9. recovery room
10. mental health room
11. therapeutic room
12. health care room
13. treatment room
14. physioth
15. recovery room
16. medical instrument room
17. restaurant
18. kitchen
19. rubbish room
20. barbershop
21. shop
22. nurse's lounge
23. duty room
24. WC

Room Area Form

function	rooms	number	unit (m²)
bedroom	single room	6	144
	double room	18	576
	suite	3	144
public room	playroom	4	272
	library	1	64
	multifunctional hall	1	128
medical services	medical room	8	204
	nurse station	3	24
management room	storehouse	6	48
	office	6	144
	staff room	2	64
subsidiary room	canteen	1	72
	kitchen	1	40
	commercial rooms	3	96

暮邻·乐居 —— 泉州生态颐养中心
LIVE WITH HAPPYINESS IN OLD AGE

Site Plan 1:1000

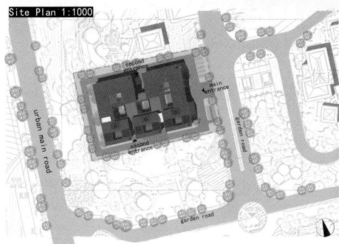

Site Analysis

Economic Indicators & Technoogy

Constrction area	3292㎡
Site area	4928㎡
Floor area ratio	0.668
Building density	0.287
Green rate	34%
Number of beds	48

West Elevation

East Elevation

The whole building ventilation analysis

Day time — Thermal Nature Ventilation

Night time — Wind Pressure Ventilation

Line Of Sight Analysis

暮邻·乐居 —— 泉州生态颐养中心
LIVE WITH HAPPYINESS IN OLD AGE

Second Floor Plan 1:200

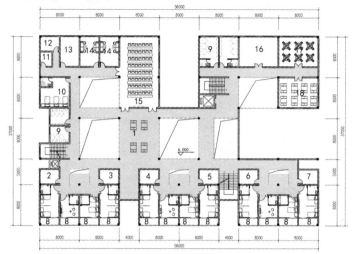

1. reading area
2. rest room
3. preparation room
4. nurse station
5. boiler room
6. store room
7. nurse's lounge
8. balcony
9. toilet
10. staff quarter
11. storehouse
12. file room
13. reception room
14. office
15. multifunctional hal
16. gymnasium
17. chess room
18. art room
19. equipment room

Third Floor Plan 1:200

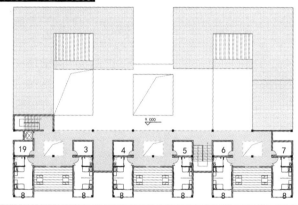

Flow Analysis

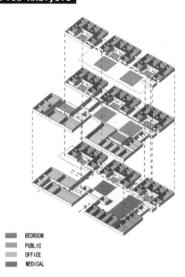

- BEDROOM
- PUBLIC
- OFFICE
- MEDICAL

Circulation Streamline

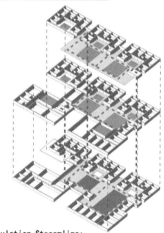

Circulation Streamline:

1. For the aged, encircles track meet their needs of walking inside. It procide the possibility for the aged to stop, watch and chat for their own preferences, once they pass difference sites.

2. For nursing assistant, it could enforce the connection among functional rooms, shorten distance and improve work efficiency.

3. For building, inside courtyard surrounded by encircled track enhance ventilation, improve light, create fine landscape environment and upgrade the living space environment.

Wall Structure

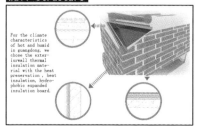

For the climate characteristics of hot and humid in guangdong, we chose the exteriorwall thermal insulation material with the heat preservation, heat insulation, hydrophobic expanded insulation board.

暮邻·乐居 ——泉州生态颐养中心
LIVE WITH HAPPYINESS IN OLD AGE

Building Structure Analysis

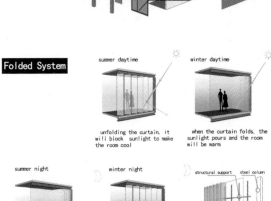

Folded System

Ventilation Analysis

Summer Daytime
During the daytime, the wind gap is open to guide the natural ventilation pass through. As the same time, the solar-house is playing an important role in sun-shading system.
When the windows is open, it can help make the air-flow in the room.

Summer Night
At summer night, wind gap of solar-house is open, the cool air will come in through the double-floor and the hot air will pass through the top windows. Meanwhile, air-handing system is transported the warm air to comfort one, accomplishing the themal cycling.

Winter Daytime
Sun-light heat the air in solar house in the south of building. Air-handing system will provide energy for the indoor, under the action of this, the warm air is transported to the north of building through the double floor, completing the thermal cycling.

Winter Night
Solar-house and air handing system absorb the heat during the daytime, release them time by time, keep the status quo to maintain the indoor temperature.

暮邻·乐居 ——泉州生态颐养中心
LIVE WITH HAPPYINESS IN OLD AGE

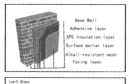
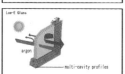
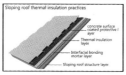
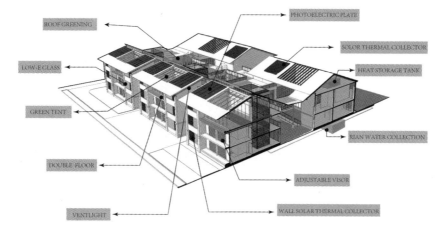
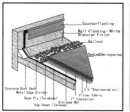
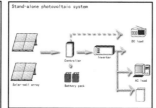
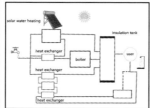
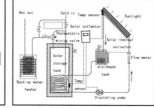

Rainwater Collection Ananlysis

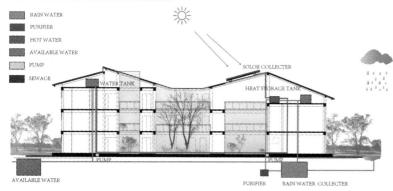

South Elevation

A-A Section

Reclaimed Water System

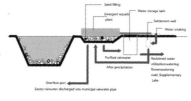

Reclaimed wastewater treatment, reclaimed water reuse in constructed wetland. The design of the water mainly includes the rain, bath water, kitchen sewage and so on. After a simple filtration, precipitation and recycling, they will be used for flushing toilets, green irrigation, road flushing and car washing and so on.

Sunshade Structure

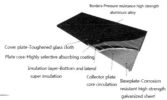

SENIOR HOME DESIGN I
A Ecological Nursing House in Quanzhou

LIGHT BOX
光·盒

综合奖·优秀奖
General Prize Awarded · Honorable Mention Prize

注 册 号：5727
项目名称：光·盒（泉州）
　　　　　Light Box（Quanzhou）
作　　者：曹鑫、关经纯、朱珊
参赛单位：南京工业大学
指导老师：蔡志昶

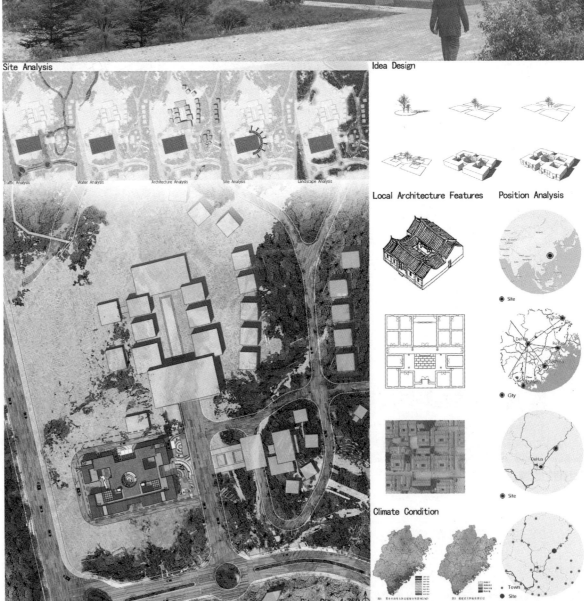

SENIOR HOME DESIGN II
A Ecological Nursing House in Quanzhou

LIGHT BOX
光·盒

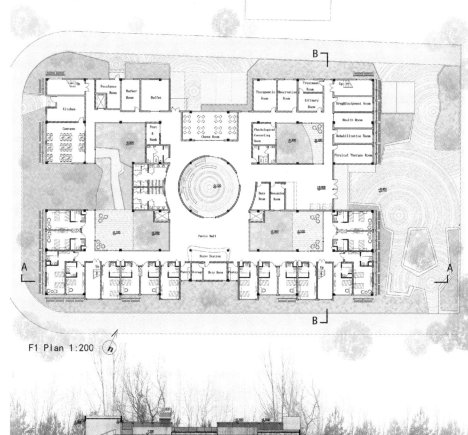

F1 Plan 1:200

Economic Analysis

Site area: 4926m²
Site coverage: 1663m²
GFA: 3298.5m²
Building density: 33.75%
Greening rate: 23.9%
FAR: 0.67
Max-height: 11.1m
Floors: 2

Design Description

This project is located in the Yinxiang ecological park, Quanzhou which is eminent for its reputation in the manufacture of ceramics. This design is targeted at accommodating a nursing house for the senior who reside in the ecological park and is combined with the park theme——ceramics. Optimizing the solar energy and the original natural condition of site, this nursing house is adapted well to the subtropical climate and has a dozen of advantages in being green, low-carbon and livable. The symmetry building is interesting in the variation of its exterior facade in virtue of the solar energy utilization components' usage. There are four courtyards inside.

The construction of the central circular courtyard is based on the abstract concept of Chinese traditional culture concerning ceramics and the theory of "a square earth and round heavens". For the sake of the concern about the senior, the building is considerate in the arrangement of barrier-free design. Extracting the essence of local traditional residences, the internal space of the building is divided into four linked groups which can eliminate the loneliness of the senior. The central courtyard is designed to accommodate social activities for the senior.

Section B-B 1:200

East Elevation 1:200

Section A-A 1:200

South Elevation 1:200

Perspective A

Perspective B

LIGHT BOX
光·盒

SENIOR HOME DESIGN III
A Ecological Nursing House in Quanzhou

F2 Plan 1:200

设计说明

本设计位于世界陶瓷之都——泉州市德化县雷峰镇瓷都印象生态园内，基地植被茂密，地表景观优美，空气清新，环境宜人。设计旨在为居住在瓷都印象生态园内的老年人提供社区养老服务。本项目结合德化瓷都印象生态园的定位，充分应用太阳能等可再生能源技术，结合周边优越的自然环境，建设了适用于中亚热带气候区的绿色、低碳、健康的生态颐养服务中心。

建筑中轴对称，结合太阳能利用构件使得造型方正而富有变化。内部共有四处庭院，中部的圆形庭院抽象了瓷器与中国传统文化中的方圆的理念。设计充分体现了对老年人生活的人性化考虑，从建筑整体考虑出发进行了无障碍设计。

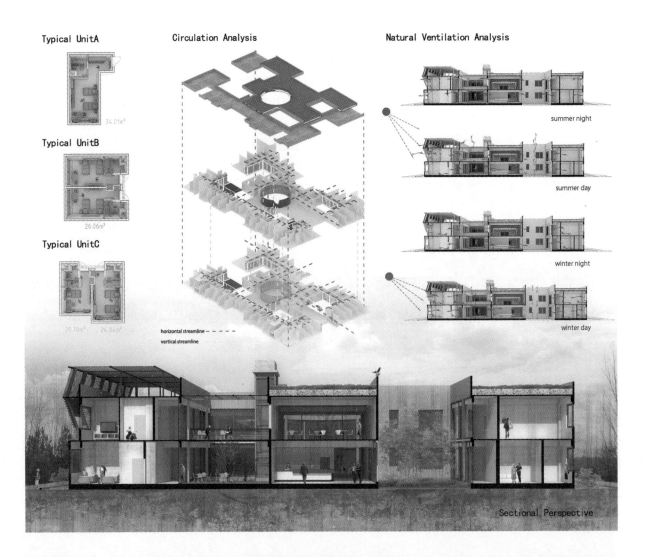

Typical UnitA

Typical UnitB

Typical UnitC

Circulation Analysis

horizontal streamline
vertical streamline

Natural Ventilation Analysis

summer night
summer day
winter night
winter day

Sectional Perspective

SENIOR HOME DESIGN IV
A Ecological Nursing House in Quanzhou

LIGHT BOX
光·盒

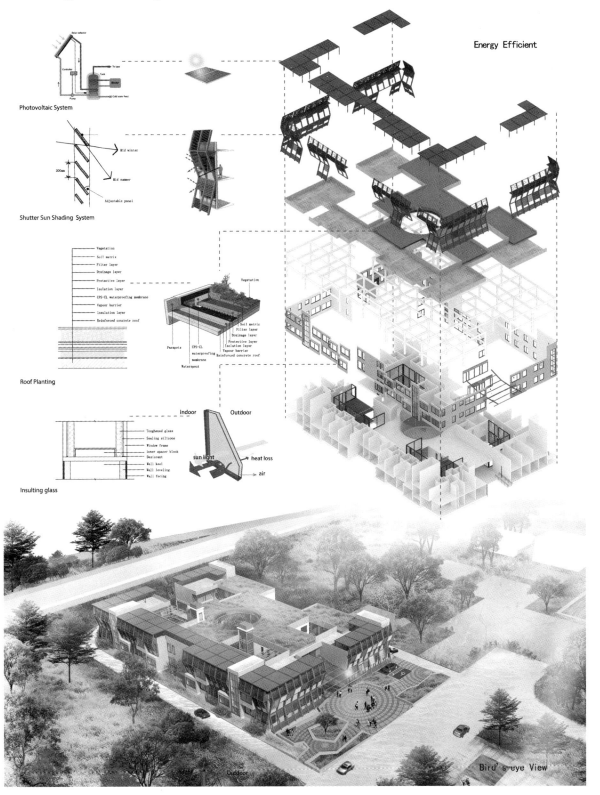

有效作品参赛团队名单
Name List of all Participants Submitting Effective Works

注册号	作者	单位名称	指导人	单位名称
4549	徐明哲、耿煜周、吴寰、赵亮	天津大学、湖南大学、深圳大学	—	—
4567	陈宗煌、郑家伟	成功大学	—	—
4596	曹祖略	Technical university of Munich	Musso Florian	Technical university of Munich
4607	袁金鹤、贾帅帅、陈浩、张弟、黄昊	安徽建筑大学	韩玲、周庆华	安徽建筑大学
4612	刘露	东北石油大学	任洪国、赵文艳、乔梁、马令勇	东北石油大学
4616	季思雨、文瑞琳、王怡璇、张艺冰、王锋宇、张瑞林	河北工业大学、石家庄铁道大学、西安建筑科技大学、中铁建安工程设计院	高力强、张军	石家庄铁道大学
4621	黄珂、吴双、李畅、胜心、刘爽	长安大学、东北石油大学	刘启波	长安大学
4624	陈冉鹏、刘炯良、王怡璇、赵冬梅、张清亮、丁磊	石家庄铁道大学、中铁建安工程设计院、天津大学	高力强、马兰	石家庄铁道大学
4633	黄方意、党昊天、杨正源	西安科技大学	孙倩倩	西安科技大学
4636	车喜刚、雷宸骁、刘雨龙、陈函璐、何静怡、成侃	西安科技大学	孙倩倩	西安科技大学
4637	王兰萍、张露露、张雁、杨吉、安建辉	西南科技大学	白雪、赵祥、董美玲	西南科技大学
4640	李晶、潘梦檀、王楚涵	西安科技大学	孙倩倩	西安科技大学
4641	黄超、陈华、张人泽	石家庄铁道大学	高力强、何国青	石家庄铁道大学
4642	刘程明、王莹莹、叶葭、全孝莉	天津大学	刘彤彤、张颀	天津大学
4645	许彦韬、杨影、文瑞琳、张艺冰、白洁媛、张清亮	石家庄铁道大学、西安建筑科技大学、中铁建安工程设计院	高力强、何国青	石家庄铁道大学
4651	郝杰、蒙胜宇	广西建设职业技术学院	—	—
4657	王长鹏、赵喆骅、张晓燕、张晓薇	山东建筑大学	张勤	山东建筑大学
4658	闫阳、吴彦强、狄岳	北方工业大学	吴正旺	北方工业大学
4662	余文杰	山西农业大学	—	—

续表

注册号	作者	单位名称	指导人	单位名称
4668	张奇、许嘉航、刘烁、陈建宇、杨正豪、冯嘉恒、李天豪、余茜	西北工业大学	陈新	西北工业大学
4669	黄一平、李停生、杨倩、柴纪阳、阚丽莹、刘奕彤、白颖	北京建筑大学	李春青、邹越	北京建筑大学
4676	黎颖枫、陆秋伶、唐麟、陈晓玲、陈俞憬	桂林理工大学	罗旋、易姗姗	桂林理工大学
4679	杨子傲、姚瑶、林昭涣、刘子玉	深圳大学	覃力	深圳大学
4681	罗欣然、苑东凯、秦昌	石家庄铁道大学	高力强、何国青	石家庄铁道大学
4683	杨丽、张丹、王泽、高子月、王艳柯、吴嘉俐	西安科技大学	孙倩倩	西安科技大学
4685	张月荣、张一鸣、袁薇、吴俣、刘澄、奥斯曼江、刘晓文、陆子骞	长安大学、西安科技大学	孙倩倩	西安科技大学
4689	高海伦、董兆、梅博涵	浙江理工大学	文强	浙江理工大学
4690	刘宜鑫、蒲宏宇、汪涟涟、汪漪漪、唐伟豪	西南科技大学、苏州大学	高明、蔡余萍、成斌	西南科技大学
4696	李垣康、冯烨、宋小玲、张塑琪、唐晓宇、李晗婧	南昌大学	—	—
4717	李习亭、李瀚威、高艺乘、李伟健	内蒙古科技大学	王文明、孙丽萍	内蒙古科技大学
4720	张云卿、胡庆强、董毓兵	石家庄铁道大学	高力强、何国青	石家庄铁道大学
4724	李舟	APM CONSULTANTS CO LTD	—	—
4729	郝俊龙、盛焱、吴新杰、张一凡、宋芳芳、孙超、樊新颖、王世博	大连理工大学	索健、陈滨	大连理工大学
4733	马云肖、徐贞、李静	西安建筑科技大学	靳亦冰、王军、毛刚	西安建筑科技大学
4737	凡开伦、李芃、岳文姬	西安建筑科技大学	—	—
4741	王长鹏、赵喆骅、韩善明	山东建筑大学	张勤	山东建筑大学
4743	贾薇、贾文学	西安建筑科技大学、秦皇岛市建设工程质量监督站	靳亦冰	西安建筑科技大学

续表

注册号	作者	单位名称	指导人	单位名称
4746	孔德皓	Yonsei university	Li sangyun	Yonsei university
4749	刘畅、李重锐、吴琼、李庆祥	天津大学	朱丽、严建伟	天津大学
4752	李少鹏、段连华、程爽	吉林建筑大学	肖景方	吉林建筑大学
4770	赵梓汐、刘梅杰	山东建筑大学	薛一冰、崔艳秋	山东建筑大学
4790	王利宇、王文伟、王冰冰、高斯如、景云峰	西安建筑科技大学	何泉、朱新荣	西安建筑科技大学
4794	陈伟、黄丽妍、刘轲	烟台大学	郑斌	烟台大学
4798	张涛、陈逸轩、罗天宇、汤钧涵	烟台大学	—	—
4802	张浩瑜、吴湛进	华中科技大学	徐燊	华中科技大学
4804	刘帆、郭梦露、田悦丰、李禹澄、曾现梦、赵晓亮	西安建筑科技大学、华侨大学	申晓辉、何泉	华侨大学、西安建筑科技大学
4805	李姝、任莹莹、王灏真、马福敬	烟台大学	郑彬	烟台大学
4817	曹煜星、张晓明、张焕、乔勇	深圳大学	陈佳伟	深圳大学
4819	沈佳成、张甘霖、高汉卿	西安建筑科技大学	—	—
4820	曹瑞华、侯俐爽、郭雷	西安建筑科技大学	—	—
4824	李劭雄、严星瑶、宋晓冉、黄锦富	东南大学	彭昌海	东南大学
4830	黄华、郑智中、李真、赵旭、徐洪光、薛芳慧、刘增军、刘冲	西安建筑科技大学	李钰、罗智星	西安建筑科技大学
4833	焦婷婷、张雷震、史明洋、王梦媛、徐成	西安建筑科技大学	何泉、朱新荣	西安建筑科技大学
4841	于沈尉、孟思佩、修泽华、朱珊、毛娅玲	河北建筑工程学院	王金奎、梅兰	河北建筑工程学院
4842	戴晨思、赵燕、张翔翔、邹清臣、袁帅	盐城工学院	黄婷、王进、徐会	盐城工学院
4843	司圣玥、陶佳燕、黄宇、朱益鑫	盐城工学院	王进、黄婷、徐会	盐城工学院
4844	唐旻雨、李子凡、师凡媛、王燕、吴昊、龚泽	盐城工学院	黄婷、王进、徐会	盐城工学院

续表

注册号	作者	单位名称	指导人	单位名称
4853	叶文杰、沈金辉、杨兴正、顾思明、吴思蓉、段涵、吴斯敏	北方工业大学	马欣、赵春喜	北方工业大学
4856	丁辛宇、丁怡文、鲍程远	吉林建筑大学	裘鞠、李雷立	吉林建筑大学
4866	汪皓、胡溪、方鹏飞、李潇然、吴海波、方柳	合肥工业大学	苏剑鸣、王旭	合肥工业大学
4867	赵子豪、王娇婧、刘畅	河北工业大学	舒平、岳晓鹏	河北工业大学
4874	宋鹏博、刁阔、吴晗、王吉亮	烟台大学	王一平	烟台大学
4879	聂大为、杨喆雨、魏婉晴、蒋子晗	大连理工大学	郭飞、于辉、苏媛	大连理工大学
4881	欧艺伟、高冠嵩、高力、王立群	湖南大学	魏春雨	湖南大学
4886	曾宪策、周涛、许可、李岩、杨昶、韩雪颖	华南理工大学	孟庆林、王扬、丘建发、宋振宇	华南理工大学
4889	景诗超、于露、袁梦、孙慧	山东建筑大学	张勤	山东建筑大学
4890	祁洋洋、巩杨帆、薛春轩、黄杨海、杨名宇、王炎初、谭雅文	河北天俱时医药化工工程设计有限公司、西北工业大学、重庆师范大学	吴农	西北工业大学
4891	李鹏飞、盛浩、乔会荣、刘雨龙	西安科技大学	裴琳娟、孙倩倩	西安科技大学
4896	张礼陶、柏陈威、申晓艺	华中科技大学	徐燊	华中科技大学
4899	陈晓彤、管再浩、王浩、曹聪	南京工业大学	胡振宇	南京工业大学
4908	高玺、王婉婷、杨思源	天津大学	郭娟利、杨葳、刘魁星、贡小雷	天津大学
4909	吕丹妮、江海华、武学志	华中科技大学	徐燊	华中科技大学
4910	王彦迪、钟毅	华中科技大学	徐燊	华中科技大学
4919	雷雯、王晓伊、韩婧、张瑞瑞	长安大学	刘启波	长安大学
4925	赵爽、张吉强、王珍珍、刘玉涵、张义新	天津大学	朱丽、高力强	天津大学
4932	彭悦、孟婕、杨鹏程、朱硕	南京工业大学	胡振宇	南京工业大学

续表

注册号	作者	单位名称	指导人	单位名称
4939	卢梦月、秦雪	华北理工大学	唐晨辉	华北理工大学
4947	黄浩、石涛、李光旭	吉林建筑大学	肖景方	吉林建筑大学
4948	高小妮、熊锋、唐超	西南科技大学	赵祥、白雪	西南科技大学
4951	冀安琪、郑方舟、杨柳青、刘鑫、傅涵葳	深圳大学	龚维敏	深圳大学
4952	刘婷、宋佳、周恩民、谢宇翔	湖南科技大学	金熙	湖南科技大学
4958	岑土沛、傅嘉伦、冯嘉瑜、张渡也、朱琪琪	深圳大学	艾志刚	深圳大学
4965	邢晔、邬昊睿	哈尔滨工业大学、贵州大学	—	—
4967	崔建东、赵婵婵、王艺晓、纪淑曼、辛媛媛	山东建筑大学	陈兴涛	山东建筑大学
4970	李瑞琪、杨艺、陶帅	北京交通大学	陈岚、杜晓辉	北京交通大学
4974	孟迪、梁玲玲	西安科技大学	孙倩倩	西安科技大学
4975	吴嘉俐、高子月、王艳柯、张丹、杨丽、王泽	西安科技大学	孙倩倩	西安科技大学
4977	郑璐阳、班鹏毅	重庆大学	周铁军、张海滨、杜晓宇	重庆大学
4981	鲁源、孙开宇、肖麒郦	西安科技大学	罗琳	西安科技大学
4982	韩平、芮万里	重庆大学	周铁军、张海滨、杜晓宇、Andrew van den Dobbelsteen	重庆大学、Delft University of Technology
4984	邱思婷、张远雪、钟星宇	西南科技大学	赵翔、高明	西南科技大学
4985	宋翠、许欣悦、于晓晨、李珏、王嘉霖	山东建筑大学	陈兴涛	山东建筑大学
4988	张军军、张宇涛、王冠军	东南大学	张宏	东南大学
4990	余莹莹、周成、王可为	浙江理工大学	邓小军	浙江理工大学
4992	王帅、周敬波、郑义	四川大学	—	—
5011	潘高、邹明妍	重庆大学	周铁军、杜晓宇、张海滨、Regina Bokel	重庆大学、Delft University of Technology
5019	刘相乾、吴杰	重庆大学	周铁军、张海滨、杜晓宇	重庆大学

续表

注册号	作者	单位名称	指导人	单位名称
5024	陆历、唐旭、杨润、杨钒	湖南科技大学	金熙	湖南科技大学
5026	姜夏、吴晓雪、李云璁	北京交通大学	李珺杰	北京交通大学
5036	崔洋、徐苇葭、沈中健	天津大学	曾坚	天津大学
5038	朱雅晖、何珊、王畅、郭嘉、沈晨思、赵夏瑀	天津大学	贾巍杨	天津大学
5041	聂子川、蔡晟霖	南昌大学	聂志勇、陈五英、周韬	南昌大学
5052	王堃、胡文斌、刘鸣飞、杨童	西安建筑科技大学	何泉、朱新荣、赵西平	西安建筑科技大学
5066	严金梦、尹冉、张晴	华北理工大学	唐晨辉	华北理工大学
5069	万华楠、毛晓天、韩赟聪	山东建筑大学	张勤	山东建筑大学
5077	杨依桓、卢衡、张颐悦、梁启阳	江苏美城建筑规划设计院有限公司、南京工业大学	靳兵、陈中宏、康景润	江苏美城建筑规划设计院有限公司、淮阴工学院
5079	姜之点、金涛、姚灵烨	南京工业大学	—	—
5102	王舟、姜睿、陈硕、徐冠宇、刘铭媛、丰晓勇	华东交通大学	陶燕	华东交通大学
5106	包翔、刘松、李永健、李敏	北京交通大学	陈岚、杜晓辉	北京交通大学
5110	吴启峰、黄新、代志宏、郭彩霞、付彤、唐前	内蒙古科技大学	白胤	内蒙古科技大学
5120	南漪、颜阿茵	重庆大学	周铁军、杜晓宇、张海滨、Andrew van den Dobbelsteen、Regina Bokel	重庆大学、Delft University of Technology
5123	温婷婷、杨皓舒、宋开鹏、王棣威	山东建筑大学	陈兴涛	山东建筑大学
5137	李胜难、李剑华、辛星、应振国	北京交通大学	杜晓辉、陈岚	北京交通大学
5142	李文娟、吕帅帅、井源泉、李炙亳、李清扬、郭路军、于雷	郑州大学	韦峰	郑州大学
5144	贾兆元、李倩芸、李磊、凌艺、郭梦铭、薛冰琳	北方工业大学	马欣、赵春喜	北方工业大学

续表

注册号	作者	单位名称	指导人	单位名称
5148	陆艺鑫、蒙宁宁、琚京蒙、张珺洁、罗小敏	北方工业大学	马欣、赵春喜	北方工业大学
5165	杨娇、陈其龙、刘颖、衣志浩	西安建筑科技大学、沈阳建筑大学	张群、赵西平	西安建筑科技大学
5167	刘玥、孟令帅、李世晨	北方工业大学、伊利诺伊理工大学	马欣、赵春喜	北方工业大学
5169	皮子臻、郭晓玲、王琳杰、赵士新、王俊然	盐城工学院	王进、黄婷、徐会	盐城工学院
5174	沈晓宇、张博闻、陈正鹏、姚灵烨	南京工业大学	胡振宇、张伟郁	南京工业大学
5176	刘爽、刘川、白莎莎	华北理工大学	—	—
5184	史家启、高骏峰、刘泽	华北理工大学	唐晨辉	华北理工大学
5186	张琮、于璐、陈萨如拉、陈孟栋	天津大学	朱丽、汪丽君	天津大学
5187	张诗婷、田春来	南京工业大学	倪震宇、张海燕	南京工业大学
5195	冉茂莎、李苏之影、丁泽源、张天祎、星梦钊、李雄飞、朱庆玲	北方工业大学	马欣、赵春喜	北方工业大学
5207	汪平平、杨小小、李晓波、侯莹莹	北京交通大学	杜晓辉、陈岚	北京交通大学
5212	杜昕睿、邓开怀、王昭	东南大学	彭昌海	东南大学
5229	李磊、主曼婷、吴鹏龙、刘姝伶、刘义鹏	合肥学院、合肥工业大学	李娟、胡毅、谢雪胜	合肥学院
5238	郭铭帅、张春华	南阳理工学院	赵敬辛	南阳理工学院
5250	侯丹蕾、陈格格、徐贺、毕雪皎、朱敦煌、罗毅、赵忆	天津大学	郭娟利、张玉坤、杨崴	天津大学
5259	曹赓、郑欣欣、林佳昕	广州大学	李丽、刘源	广州大学
5266	赵姝雅、张晶玫、王治斌、陈煜珩	中国矿业大学（北京）	谢略	中国矿业大学（北京）
5271	黄皇、万奕奇、郑智文	深圳大学	杨文焱	深圳大学
5274	刘心怡、张晶、李炳云、朱文景、张锐	山东建筑大学	张勤	山东建筑大学
5275	刘莎莎、马凯、孙文祺、卫锋、李胤恺	山东建筑大学	张勤	山东建筑大学
5276	罗多、余国保、邓鑫、莫日根、赵玉龙、詹畅、徐振玉、杨少刚、邱超、李建莹、路欢琪、陈进文、宋满香、曾轶、徐飞宇	珠海兴业绿色建筑科技有限公司	—	—

续表

注册号	作者	单位名称	指导人	单位名称
5283	王楠、王劲柳	天津大学	刘丛红、杨鸿玮	天津大学
5292	王开蕊、苟堂、陈希、王丛越、吉俊杰	重庆大学	周铁军、宗德新、杜晓宇、张海滨	重庆大学
5295	高旭	本匠营建筑设计工作室	—	—
5299	范琳琳、吴则鸣	东南大学	杨维菊、吴锦绣	东南大学
5302	冯亚茜、徐意敏	山东科技大学	冯巍	山东科技大学
5309	王丛越、吉俊杰、王开蕊、苟堂	重庆大学	周铁军、杜晓宇、张海滨	重庆大学
5310	曾家敏、李拓	东北石油大学	李静薇	东北石油大学
5314	陈文茶	同济大学浙江学院	邓靖	同济大学浙江学院
5317	闫颖超、朱冉睿、侯可阳	深圳大学	赵勇伟	深圳大学
5318	杨博	WY工作室	—	—
5331	曹鑫鑫	天津大学	赵建波、宋昆	天津大学
5332	杨琦辉、曾靖云、郑斯闻、王玉琳、王波、涂小锵、杨琪、柯兰晶	华侨大学	冉茂宇、申晓辉、薛佳薇	华侨大学
5336	初堃、叶成萌、刘秋实	华北理工大学	唐晨辉、屈志刚	华北理工大学
5346	郑卓凡、杜银洁	同济大学浙江学院	Eva、陈卓	同济大学浙江学院
5347	薛诗佳	同济大学浙江学院	邓靖	同济大学浙江学院
5349	高俊峰、王亚鑫、王堃	内蒙古科技大学	陈冰俐、魏融、乌进高娃	内蒙古科技大学
5359	张菲菲、王欣、安琪、李洁	山东建筑大学	张勤、薛一冰、周中凯	山东建筑大学
5361	汪海鑫、阿里木、金云、苏昱	大连理工大学	胡文荟	大连理工大学
5364	范杰玙、许灿霖、王嘉鹏	同济大学浙江学院	邓靖	同济大学浙江学院
5367	刘慧敏、周宗宇	重庆大学	周铁军、张海滨、杜晓宇、Bokel Regina、Andy van den Dobbelsteen、Michiel Fremouw	重庆大学、Delft University of Technology

续表

注册号	作者	单位名称	指导人	单位名称
5369	汤国威、胡培圣、毛潘、贾晓伟、万怡、葛碧秋、王新苗、吕蒙、陈卓辰、田寅山	西安建筑科技大学	何泉	西安建筑科技大学
5386	张鹏、黄晶晶、黎家雄、邹清妍	深圳大学	仲德崑、龚维敏	深圳大学
5388	张煜、陈荣华、吕颖洁、吴江源、范钦锋	东南大学	杨维菊	东南大学
5393	徐峥晖、田正涛	南阳师范学院	高蕾	南阳师范学院
5404	郎冰、刘刚、李思萌	山东建筑大学	崔艳秋、薛一冰	山东建筑大学
5411	查翔、高子修、蒋晨晨、崔项、陈烨、郑兴宇	南京工业大学	张海燕	南京工业大学
5412	孙杰、罗梓豪、卜娴慧	广东省建筑设计研究院东莞分院	—	—
5413	王敏舟	同济大学浙江学院	邓靖	同济大学浙江学院
5414	满雯佳、朱华、李东和、王敏	山东建筑大学	张勤	山东建筑大学
5417	蒋星宇、张楠、吕欣宇、王洪彬	盐城工学院	王进、黄婷、徐会	盐城工学院
5423	张少飞、曾于舒、杨笛、朱永新、李奇芫	天津大学	李伟	天津大学
5426	杨静熠、许佳慧、杨智隆	同济大学浙江学院	邓靖	同济大学浙江学院
5429	蒋泽智、杨巧状、陈永臻、李新博、胡星	四川农业大学	张丽丽	四川农业大学
5432	镡旭璐、陈功勤	厦门大学	王绍森	厦门大学
5434	陶鹤友、徐骁斌、胡婷婷	东南大学	杨维菊、袁玮、万邦伟	东南大学、东南大学建筑设计研究院
5437	徐一凡、俞恒	同济大学浙江学院	Eva、陈卓	同济大学浙江学院
5440	张玺、倪晨辉、孙宁晗、张子奇、万华楠、毛晓天、韩赟聪	山东建筑大学、北京建筑大学	赵学义、崔艳秋、欧阳文	山东建筑大学、北京建筑大学
5441	刘轩、孟少佳、郑婷、张坚、刘江	西安建筑科技大学	马纯立、何泉、朱新荣、罗智星	西安建筑科技大学
5444	宋祥中、宗烨、宋子叶、寇荆坡	山东建筑大学	张勤	山东建筑大学
5449	周韦博	University of Texas at Austin	Uil Dangel	University of Texas at Austin

续表

注册号	作者	单位名称	指导人	单位名称
5452	李阔、王芷菲、郭腾	沈阳建筑大学、济南大学	黄勇	沈阳建筑大学
5454	陈忠芳、张星月、夏钰婷、张宇晗、赵川	西南科技大学	赵祥、白雪	西南科技大学
5455	赵婕、潘钰琰、劳妍婧	同济大学浙江学院	陈卓、Eva	同济大学浙江学院
5458	徐笑、白梅、潘莹莹、钟江龙	华中科技大学	—	—
5460	白昕洲、任若思、涂高雅、李明硕、马梦如	河北工程大学	王晓健、霍玉佼	河北工程大学
5461	郭嘉钰、刘家韦华、郭璇、时雯	河北工程大学	田芳、霍玉佼	河北工程大学
5462	王斌、王怡春、聂琦、辛颖	华南理工大学	王璐、王静	华南理工大学
5471	蔡希鹏、曹云琥	南京工业大学	邵继中	南京工业大学
5474	孔圣丹、石伊萱	南京工业大学	邵继中	南京工业大学
5475	耿思雨、高雯、刘鼎艺、王光宇	北京交通大学	陈泳全、曾忠忠	北京交通大学
5476	许航彬	同济大学浙江学院	邓靖	同济大学浙江学院
5480	陈培林、黄学海、唐浩铭、李琛、王垚霓、刘欣雨	西南科技大学	赵祥、白雪	西南科技大学
5487	王晗星、王湛	同济大学浙江学院	陈卓	同济大学浙江学院
5491	姜云瀚、刘焕杰、佟岩、赵晓璇、王月悦	山东科技大学	周同、谢栋、夏斐	山东科技大学
5509	王晶、宋咏洁、张端阳、王汉斌、李梓维、张洺瑜、郭璐杰、李艳雯	中国矿业大学（北京）	李晓丹	中国矿业大学（北京）
5510	魏雪仪、周妍睿、余睿、王鹏飞、蒋亮	四川农业大学	张丽丽、侯超平	四川农业大学
5513	张立、高梦雨、袁梦、于露、张蒙	山东建筑大学	张勤、薛一冰	山东建筑大学
5517	张洺瑜、郭璐杰、李艳雯、张端阳、王晶、宋咏洁、王汉斌、李梓维	中国矿业大学（北京）	李晓丹	中国矿业大学（北京）
5523	范钦锋、张维鹏、于佳欣、张煜	东南大学、汉能控股集团	杨维菊	东南大学
5524	邓佳滢、郑于恬、曾静瑶	广州大学	李丽、刘源	广州大学
5525	齐一泓、王旭楠、闫岩、才俊	沈阳建筑大学	黄勇、王飒	沈阳建筑大学

续表

注册号	作者	单位名称	指导人	单位名称
5527	杨阳、楚家麟、吴晨	沈阳建筑大学	黄勇	沈阳建筑大学
5528	朱星哲、徐冰清、李彩霞、刘培佩、邱晓婧	厦门大学	凌世德、张燕来、王伟	厦门大学
5529	幸彬、谢欣、王思远、裴朋雨、黄键、梁启基	广州大学	李丽、刘源	广州大学
5545	梁景怡、黄丽娟、何嘉惠、张敏行、刘钰暄	广州大学	李丽、刘源	广州大学
5555	王忠义、甘羽	合肥工业大学	宣晓东	合肥工业大学
5556	李曼婷、岳凯	南阳理工学院	郑方圆	南阳理工学院
5572	向柃蒨、岳婷	华中科技大学	徐燊	华中科技大学
5574	宋如意、张路薇	南京工业大学	张海燕、陈晓扬、倪震宇	南京工业大学、东南大学
5575	黄麒龙、叶伦源、刘飞	西南科技大学	赵祥	西南科技大学
5582	刘浔风、胡正元、鲍华英、汪若曦、朱炳哲、王欣、冯宏欣、林培旭	天津大学	郭娟丽、杨崴	天津大学
5584	吴浩然、徐墨林、吴宽、邵丹	天津大学、南开大学	张玉坤、赵劲松	天津大学
5585	陆垠、顾晨	南京工业大学	张海燕	南京工业大学
5587	裴婉煜、李靖、胡平、张雅琳、杜衍旭、郭嘉钰	河北工程大学	王晓健、霍玉佼	河北工程大学
5589	陈汐、张蕙麟、龚然	西南交通大学	—	—
5596	徐益翔、罗景文、任静	西南科技大学	赵祥	西南科技大学
5605	王博询、朱哲炫、林清锡、宣屹颖、胡亚芬	中原工学院、浙江工业大学之江学院、武汉大学	王宇洁	浙江工业大学之江学院
5607	宋智霖、宣勤朗、戴安李、孙力枰、周立	厦门大学	罗林	厦门大学
5611	张舒然、梁宇珅、李奇芫、孙鹏雄、赵曼	天津大学	张玉坤、张睿	天津大学
5615	颜子阳、皮开杰、李浩南	西南科技大学	赵祥、刘繁春	西南科技大学
5620	廖伟平、朱清尘、谢博颖	北京交通大学	杜晓辉、王鑫	北京交通大学

续表

注册号	作者	单位名称	指导人	单位名称
5627	水若晨、徐嘉栋、汪颖	南京工业大学	胡振宇	南京工业大学
5636	李梦楠、刘超成	南京工业大学	张海燕	南京工业大学
5642	彭卓、袁思奇、杨肖、刘高阳、吴玮琳	南昌大学	—	—
5643	刘烨、王鑫	北京建筑大学、北京交通大学	—	—
5647	罗开洲、刘国军	南阳师范学院	高蕾	南阳师范学院
5648	张浩楠、刘敏、邓小薇、李霞	华中科技大学	刘剀	华中科技大学
5654	韩梦、魏肖翔、秦朝晖、张昊、晋露文、屠悦	西安建筑科技大学	张倩、王芳	西安建筑科技大学
5677	温良涵、王涵、王浴安	大连理工大学	—	—
5688	表秀峰、金姗	沈阳建筑大学	李勇	沈阳建筑大学
5692	苏冰霜	南阳师范学院	高蕾	南阳师范学院
5695	王雨寒、谭宇星、邬曹闽、谢嘉荷、熊依晴、冯思达、支佳钰、李珂	南昌大学	徐丛淮	南昌大学
5696	陈艺松、刘来凤、赵航宇、岳开云	石家庄铁道大学	高力强、何国青	石家庄铁道大学
5707	吴容蓉、冉佳珞、赵方圆、杨鸿基、徐宝伟	重庆大学	李百毅、黄海静	重庆大学
5725	杨云帆、王雨晨	南京工业大学	蔡志昶	南京工业大学
5727	曹鑫、关经纯、朱珊	南京工业大学	蔡志昶	南京工业大学

2017台达杯国际太阳能建筑设计竞赛办法
Competition Brief for International Solar Building Design Competition 2017

竞赛宗旨：

银龄化背景下，每个人都关注在何种建筑中、以何种方式安度晚年。通过优化建筑设计手段，整合适宜的可再生能源技术，让太阳能为建筑带来永续的能源，让建筑为老人创造安全、健康、舒适、便利、绿色的新生活。

竞赛主题：阳光·颐养

竞赛题目：西安生态颐养服务中心项目
　　　　　泉州生态颐养服务中心项目

主办单位：国际太阳能学会
　　　　　中国可再生能源学会
　　　　　全国高等学校建筑学学科专业指导委员会

承办单位：国家住宅与居住环境工程技术研究中心
　　　　　中国可再生能源学会太阳能建筑专业委员会

协办单位：中国建筑设计院有限公司

冠名单位：台达环境与教育基金会

媒体支持：《建筑技艺》(AT)杂志

评委会专家：崔愷：国际建筑师协会副理事、中国建筑学会常务理事、中国工程院院士、国家工程设计大师、中国建筑设计院有限公司总建筑师。
Deo Prasad：国际太阳能学会亚太区主席，澳大利亚新南威尔士大学建筑环境系教授。
M.Norbert Fisch：德国不伦瑞克理工大学教授（TU Braunschweig），建筑与太阳能技术学院院长。

GOAL OF COMPETITION:

Under the background of aging, everyone is paying attention to what kind of building they will live in, and how to spend their twilight years. Through optimizing architectural design means and integrating appropriate renewable energy technology, solar energy can bring sustainable energy for buildings and the buildings can create a safe, healthy, comfortable, convenient, green new life for the old.

THEME OF COMPETITION:

SUNSHINE AND CARE FOR THE ELDERLY

SUBJECT OF COMPETITION:

Subject I: Ecological Elderly Care Service Center in Xi'an City
Subject II: Ecological Elderly Care Service Center in Quanzhou City

HOST:

International Solar Energy Society
Chinese Renewable Energy Society
National Supervision Board of Architectural Education (China)

ORGANIZER:

China National Engineering Research Center for Human Settlements
China Renewable Energy Society solar building Specialized Committee

COORGANIZER:

China Architectural Design Institute Co., Ltd.

Peter Luscuere：荷兰代尔伏特大学(TU Delft)建筑系教授。
Mitsuhiro Udagawa：国际太阳能学会日本区主席，日本工学院大学建筑系教授。
杨经文：马来西亚汉沙杨建筑师事务所创始人。
林宪德：台湾绿色建筑委员会主席，台湾成功大学建筑系教授。
仲继寿：中国可再生能源学会太阳能建筑专业委员会主任委员，国家住宅与居住环境工程技术研究中心主任。
喜文华：联合国工业发展组织国际太阳能技术促进转让中心主任，联合国可再生能源国际专家，国际协调员，甘肃自然能源研究所所长。
黄秋平：华东建筑设计研究总院副总建筑师。
冯雅：中国建筑西南设计研究院副总工程师，中国建筑学会建筑热工与节能专业委员会副主任。

组委会成员：由主办单位、承办单位及冠名单位相关人员组成。办事机构设在中国可再生能源学会太阳能建筑专业委员会。

评比办法：

1. 由组委会审查参赛资格，并确定入围作品。
2. 由评委会评选出竞赛获奖作品。

评比标准：

1. 参赛作品须符合本竞赛"作品要求"的内容。
2. 作品应具有原创性和前瞻性，鼓励创新。
3. 作品应满足使用功能、绿色低碳、安全健康的要求，建筑技术与太阳能利用技术具有适配性。
4. 作品应充分体现太阳能利用技术对降低建筑使用能耗的作用，在经济、技术层面具有可实施性。
5. 作品评定采用百分制，分项分值见下表：

TITLE SPONSOR:

Delta environment and Education Foundation

MEDIA SUPPORT:

Architecture Technique

EXPERTS OF JUDGING PANEL:

Mr. Cui Kai, Academician of China Academy of Engineering, Deputy Board Member of IUA (International Union of Architects); Vice President of Architectural Society of China; National Design Master and Chief Architect of China Architecture Design & Research Group.

Mr. Deo Prasad, Asia-Pacific President of International Solar Energy Society (ISES) and Professor of Faculty of the Built Environment, University of New South Wales, Sydney, Australia.

Mr. M. NorbertFisch, Professor of TU Braunschweig and president of the Institute of Architecture and Solar Energy Technology, Germany.

Mr. Peter Luscuere, Professor of Department of Architecture, Delft University of Technology, The Netherlands.

Mr. Mitsuhiro Udagawa, President of ISES-Japan and professor of Department of Architecture, KogakuinUniversity.

Mr. Yang Jingwen：Principal of T. R. Hamzah & Yeang Sdn. Bhd. (Malaysia).

Mr. Lin Xiande, President of Taiwan Green Building Committee and Professor of Faculty of Architecture of Success University, Taiwan.

Mr. Zhong Jishou, Chief Commissioner of Special Committee of Solar Buildings, CRES and Director of CNERCHS.

Mr. Xi Wenhua, Director-General of Gansu Natural Energy Research Institute; Director-General of UNIDO International Solar Energy Center for Technology Promotion and Transfer; expert in sustainable energy field from United Nations, international coordinator; Director, Gansu Institute of natural energy.

Mr. Huang Qiuping: Deputy General Architect of Huadong Institute of architectural esign and Research.

Mr. Feng Ya, Deputy chief engineer of Southwest Architecture Design and Research Institute of China; deputy director of special committee of building thermal and energy efficiency, Architectural Society of China.

MEMBERS OF THE ORGANIZING COMMITTEE:

It is composed by competition organizer, operator and sponsor. The administration office is a standing body in Special Committee of Solar Buildings, CRES

APPRAISAL METHODS:

1. Organizing Committee will check up eligible entries and confirm shortlist entries.

评比指标	指标说明	分值
规划与建筑设计	规划布局、环境利用与融入、功能流线、无障碍设计、建筑艺术，鼓励创新	40
被动太阳能利用技术	通过专门建筑设计与建筑构造利用太阳能的技术，鼓励创新	30
主动太阳能利用技术	通过专门设备收集、转换、传输、利用太阳能的技术，鼓励创新	10
采用的其他技术	其他绿色、低碳、安全、健康技术，鼓励创新	10
技术的可操作性	作品的可实施性，技术的经济性和普适性要求	10

设计任务书及专业术语等附件：（见资料下载）

附件1：西安生态颐养服务中心项目
附件2：泉州生态颐养服务中心项目
附件3：专业术语
附件4：参赛者信息表
附件5：作品设计方法报告

奖项设置及奖励形式：

1. 综合奖：
一等奖作品：2名　颁发奖杯、证书及人民币50000元奖金（税前）；
二等奖作品：4名　颁发奖杯、证书及人民币20000元奖金（税前）；
三等奖作品：6名　颁发奖杯、证书及人民币5000元奖金（税前）；
优秀奖作品：30名　颁发证书。
2. 优秀设计方法奖：10名　颁发证书及人民币2000元奖金（税前）。
作品设计方法报告内容丰富充实，设计过程记录完整，设计方法新颖。
3. 技术专项奖：名额不限，颁发证书。
作品采用的技术或设计方面具有创新，实用性强。
4. 建筑创意奖：名额不限，颁发证书。
作品规划及建筑设计方面具有独特创意和先导性。

2.Judging Panel will appraise and select out the awarded works.

APPRAISAL STANDARD:

1. The entries must meet the demands of the Competition Requirement.
2. The entries should embody originality and prospective in order to encourage innovation.
3. The submission works should meet the demands of usable function, green and low-carbon, and health and coziness. The building technology and solar energy technology should have adaptability to each other.
4. The submission works should play the role of reducing building energy consumption by utilization of solar energy technology and have feasibility in the aspect of economy and technology.
5. A percentile score system is adopted for the appraisal as follows:

APPRAISAL INDICATOR	EXPLANATION	SCORES
Planning and Architecture design	Urban planning design, use of environmental resource and integrating into the surroundings, functional division and streamline organization, barrier free design, architectural art. Innovation is encouraged	40
Utilization of passive solar energy technology	Use of solar energy by specific architecture and construction design. Innovation is encouraged	30
Utilization of active solar energy technology	Use of solar energy though collecting, transforming, and transmitting energy by specific equipment. Innovation is encouraged	10
Other technologies	Other technologies such as: green, low carbon, safe and healthy technologies. Innovation is encouraged	10
Operability of the technology	Feasibility, economy, and popularity of relevant technology demands	10

THE TASK BOOK OFDESIGN AND PROFESSIONAL GLOSSARY (Found in Annex)

Annex 1: Ecological Elderly Care Service Center in Xi'an City
Annex 2: Ecological Elderly Care Service Center in Quanzhou City
Annex 3: Professional Glossary
Annex 4: Information table
Annex 5: Report of works' design method

作品要求：

1. 建筑设计方面应达到方案设计深度，技术应用方面应有相关的技术图纸和指标。

2. 作品图面、文字表达清楚，数据准确。

3. 作品基本内容包括：

3.1 简要建筑方案设计说明（限200字以内），包括方案构思、太阳能综合应用技术与设计创新、技术经济指标表等。

3.2 项目的竞赛作品需进行竞赛用地范围内的规划设计，总平面图比例为 1：500～1：1000（含活动场地及环境设计）。

3.3 单体设计：

能充分表达建筑与室内外环境关系的各层平面图、外立面图、剖面图，比例 1：200；

能表现出技术与建筑结合的重点部位、局部详图及节点大样，比例自定；其他相关的技术图、分析图、表等。

3.4 建筑效果表现图1～4个。

3.5 参赛者须将作品文件编排在 840mm×590mm 的展板区域内（统一采用竖向构图），作品张数应为4或6张。中英文统一使用黑体字。字体大小应符合下列要求：标题字高：25mm；一级标题字高：20mm；二级标题字高：15mm；图名字高：10mm；中文设计说明字高：8mm；英文设计说明字高：6mm；尺寸及标注字高：6mm。文件分辨率100dpi，格式为JPG或PDF文件。

4. 参赛者通过竞赛网页上传功能将作品递交竞赛组委会，入围作品由组委会统一编辑板眉、出图、制作展板。

5. 作品文字要求：除3.1"建筑方案设计说明"采用中英文外，其他为英文；建议使用附件3中提供的专业术语。

参赛要求：

1. 欢迎建筑设计院、高等院校、研究机构、绿色建筑部品研发生产企业等单位，组织专业人员组成竞赛小组参加竞赛。

2. 请参赛者访问 www.isbdc.cn，按照规定步骤填写注册表，提交后会得到唯一的注册号，即为作品编号，一个作品对应一个注册号。提交作品时把注册号标注在每副作品的左上角，字高6mm。注册时间2016年6月1日～2016年12月31日。

3. 参赛者同意组委会公开刊登、出版、展览、应用其作品。

Award Setting and Awards Form:

1. GENERAL PRIZES:

First Prize: 2 winners
The Trophy Cup, Certificate and Bonus RMB 50,000 (before tax) will be awarded.

Second Prize: 4 winners
The Trophy Cup, Certificate and Bonus RMB 20,000 (before tax) will be awarded.

Third Prize: 6 winners
The Trophy Cup, Certificate and Bonus RMB 5,000 (before tax) will be awarded.

Honorable Mention Prize: 30 winners
The Certificate will be awarded.

2. PRIZE FOR EXCELLENT DESIGN METHOD: 10 winners
The Certificate and Bonus RMB 2000 (before tax) will be awarded.
The report of works' design method is rich and full, the design process is complete, and the design method is novel.

3. PRIZE FOR TECHNICAL EXCELLENCE WORKS:
The quota is open-ended. The Certificate will be awarded.
Prize works must be innovative with practicability in aspect of technology adopted or design.

4. PRIZE FOR ARCHITECTURAL ORIGINALITY:
The quota is open-ended. The Certificate will be awarded.
Prize works must be originally creative and advanced.

REQIREMENTS OF THE WORK:

1. The submitted drawing sheets should meet the requirements of scheme design level and should be accompanied with relevant technical drawings and technology data.

2. Drawings and text should be expressed in clear and readable way. Mentioned data should be accurate.

3. The submitted work should include:

3.1 A project description (not exceeding 200 words) including the following factors: Schematic concept design description; Integration of solar energy technology; Innovative design; Technical and economic indicators.

3.2 Participants should provide an urban design within the outline of the site of the competition. Participants will provide a site plan (including urban context / urban design) with the scale of 1：500 or 1：1000.

3.3 Monomer Design:

Participants will provide floor plans, elevations and sections with the scale of 1：200, which can fully express the relationship between architecture and indoor and outdoor environment

Participants should provide detailed drawings (without limitation of scale) that illustrate the integration of technology in the architectural project, as well as any other relevant elements, such as technical charts, analysis diagram, and tables.

3.4 Rendering perspective drawing (1~4).

3.5 Participants should arrange the submission into four or six exhibition

4. 被编入获奖作品集的作者，应配合组委会，按照出版要求对作品进行相应调整。

5. 参赛者需提交作品设计方法报告，见附件5。

注意事项：

1. 参赛作品电子文档和作品设计方法报告须在2017年3月1日前提交组委会，请参赛者访问www.isbdc.cn，并上传文件，不接受其他递交方式。

2. 作品中不能出现任何与作者信息有关的标记内容，否则将视其为无效作品。

3. 组委会将及时在网上公布入选结果及评比情况，将获奖作品整理出版，并对获奖者予以表彰和奖励。

4. 获奖作品集首次出版后30日内，组委会向获奖作品的创作团队赠样书2册。

5. 竞赛活动消息发布、竞赛问题解答均可登陆竞赛网站查询。

所有权及版权声明：

参赛者提交作品之前，请详细阅读以下条款，充分理解并表示同意。

依据中国有关法律法规，凡主动提交作品的"参赛者"或"作者"，主办方认为其已经对所提交的作品版权归属作如下不可撤销声明：

1. 原创声明

参赛作品是参赛者原创作品，未侵犯任何他人的任何专利、著作权、商标权及其他知识产权；该作品未在报纸、杂志、网站及其他媒体公开发表，未申请专利或进行版权登记，未参加过其他比赛，未以任何形式进入商业渠道。参赛者保证参赛作品终身不以同一作品形式参加其他的设计比赛或转让给他方。否则，主办单位将取消其参赛、入围与获奖资格，收回奖金、奖品及并保留追究法律责任的权利。

2. 参赛作品知识产权归属

为了更广泛推广竞赛成果，所有参赛作品除作者署名权以外的全部著作权归竞赛承办单位及冠名单位所有，包括但不限于以下方式行使著作权：享有对所属竞赛作品方案进行再设计、生产、销售、展示、出版和宣传的权利；享有自行使用、授权他人使用参赛作品用于实地建设的权利。竞赛主办方对所有参赛作品拥有展示和宣传等权利。其他任何单位和个人（包括参赛者本人）未经授权不得以任何

panels, each 840mm × 590mm in size (arranged vertically). Chinese and English font type should be both in boldface. Font height is required as follows: title with word height 25mm; first subtitle with word height 20mm; second subtitle: word height 15mm; figure title: word height 10 mm; design description word height 6mm; dimensions and labels: 6mm. File resolution: 100dpi in JPEG or PDF format.

4. Participants should send (upload) a digital version of submission via FTP to the organizing committee, who will compile, print and make exhibition panels for shortlist works.

5. Text requirement: The submission should be in English, in addition to 3.1 "architectural design description" in English and Chinese. Participants should use the words from the Professional Glossary in Appendix 3.

PARTICIPATION REQUIREMENTS:

1. Institutes of architectural design, colleges and universities, research institutions and green building product development and manufacturing enterprises organize professionals are welcomed to form a competition group to take part in the competition.

2. Please visit www.isbdc.cn. You may fill the registry according to the instruction and gain an ID of your work after submitting the registry, that is work number. One work only has one ID. The number should be indicated in the top left corner of each submission work with word height in 6mm. Registration time: 1st June, 2016 – 31st December, 2016.

3. Participants must agree that the Organizing Committee may publish, print, exhibit and apply their works in public.

4. The authors whose works are edited into the publication should cooperate with the Organizing Committee to adjust their works according to the requirements of press.

5. Participants are required to submit a Report of works' design method, which will be found in Annex 5.

IMPORTANT CONSIDERATION:

1. Participant's digital file and works' design method report must be uploaded to the organizing committee's FTP site (www.isbdc.cn) before 1st March, 2017. Other ways will not be accepted.

2. Any mark, sign or name related to participant's identity should not appear in, on or included with submission files, otherwise the submission will be deemed invalid.

3. The Organizing Committee will publicize the process and result of the appraisal online in a timely manner, compile and publish the awarded works. The winners will be honored and awarded.

4. In 30 days after the collection of works being published, 2 books of award works will be freely presented by the Organizing Committee to the competition teams who are awarded.

5. The information concerning the competition as well as explanation about all activities may be checked and inquired in the website of the competition.

形式对作品转让、复制、转载、传播、摘编、出版、发行、许可使用等。参赛者同意竞赛承办单位及冠名单位在使用参赛作品时将对其作者予以署名，同时对作品将按出版或建设的要求作技术性处理。参赛作品均不退还。

3. 参赛者应对所提交作品的著作权承担责任，凡由于参赛作品而引发的著作权属纠纷均应由作者本人负责。

声明：

1. 参与本次竞赛的活动各方（包括参赛者、评委和组委），即表明已接受上述要求。

2. 本次竞赛的参赛者，须接受评委会的评审决定作为最终竞赛结果。

3. 组委会对竞赛活动具有最终的解释权。

4. 为维护参赛者的合法权益，主办方特请参赛者对本办法的全部条款，特别是"所有权及版权"声明部分予以充分注意。

附件1：
西安生态颐养服务中心项目

一、西安生态颐养服务中心项目气候条件

项目用地位于西安市世界地质公园秦岭山脉终南山下子午镇台沟村，北纬34°02′，东经108°53′，处于渭河流域中部关中盆地，最低海拔384.7m，最高海拔2886.7m。地势南高北低，东高西低，南为秦岭山地，北为渭河断陷谷地冲积平原区（包括台原），西为渭河冲积平原（含秦岭北麓洪积扇群），东部为黄土台原与川道沟壑。

西安市长安区属于暖温带半湿润大陆性季风气候区，雨量适中，四季分明，气候温和，秋短春长。冬季比较干燥寒冷，春季温暖，夏季炎热多雨，秋季温和湿润。年平均气温13.3℃，年均降水量654mm，湿度69.6%，日照1377h。最冷的1月份平均气温0.3℃，最热的7月份平均气温26.9℃。全年多东北风，年平均风速为1.3～2.6m/s。长安地热资源丰富，地热面积约207.5km^2，热能储量相当于1324.5万吨标准煤的热能。

基本气象资料

气象参数：北纬（34°18′）东经（108°55′）、测量点海拔高度398m。

ANNOUNCEMENT ABOUT OWNERSHIP AND COPYRIGHT:

Before submitting the works, participants should carefully read following clauses, fully understand and agree with them.

According to relevant national laws and codes it is made sure by the competition sponsors that all "participants" or "authors" who have submitted their works on their own initiative have received following irrevocable announcement concerning the ownership of their works submitted:

1. Announcement of originality

The entry work of the participant is original, which does not infringe any patent, copyright, trademark and other intellectual property; it has not been published in any newspapers, periodicals, magazines, webs or other media, has not been applied for any patent or copyright, not been involved in any other competition, and not been put in any commercial channels. The participant should assure that the work has not been put in any other competition by the same work form in its whole life or legally transferred to others, otherwise, the competition sponsors will cancel the qualification of participation, being shortlisted and awarded of the participant, call back the prize and award and reserve the right of legal liability.

2. The ownership of intellectual property of the works

In order to promote competition results, the participants should relinquish copyright of all works to competition administrators and titled unit except authorship. It includes but is not limited to the exercise of copyright as follows: benefit from the right of the works on redesigning, production, selling, exhibition, publishing and publicity; benefit from the right of the works on construction for self use or accrediting to others for use. Without accreditation any organizations and individual (including authors themselves) cannot transfer, copy, reprint, promulgate, extract and edit, publish and admit to use the works by any way. Participants have to agree that competition administrators and titled unit will sign the name of authors when their works are used and the works will be treated for technical processing according to the requirements of publication and construction. All works are not returned to the author.

3. All authors must take responsibility for their copyrights of the works including all disputes of copyright caused by the works

ANNOUNCEMENT:

1. It implies that everybody who has attended the competition activities including participants, jury members and members of the Organizing Committee has accepted all requirements mentioned above.

2. All participants must accept the appraisal of the jury as the final result of the competition.

3. The Organizing Committee reserves final right to interpret for the competition activities.

4. In order to safeguard the legitimate rights and interests of the participants, the organizers ask participants to fully pay attention to all clauses in this document, especially some clauses with blue colors.

月份	空气温度 °C	相对湿度 %	水平面日太阳辐射 kW·h/(m²·d)	大气压力 kPa	风速 m/s	土地温度 °C	月采暖度日数 °C·d	供冷度日数 °C·d
一月	0.3	61.4	2.2	89.8	1.5	-4.62	549	0
二月	3.7	57.9	2.64	89.7	2.0	-0.82	400	0
三月	8.6	62.3	3.22	89.4	2.2	5.0	291	0
四月	15.3	64.4	4.34	89.1	2.2	13.2	81	159
五月	20.3	66.1	4.04	88.9	2.3	18.9	0	319
六月	25.1	61.5	4.5	88.5	2.2	22.8	0	453
七月	26.9	69.7	4.16	88.4	2.3	24.2	0	524
八月	25.1	76.4	4.85	88.7	2.3	22.5	0	468
九月	20.4	77.0	3.08	89.3	1.9	17.8	0	312
十月	14.3	76.3	3.17	89.7	1.5	11.1	115	133
十一月	7.2	71.5	2.09	89.9	1.5	4.1	324	0
十二月	1.4	65.6	1.75	90	1.4	-2.2	515	0
年平均数	14.1	67.6	3.34	89.3	1.9	11.0	2275	2369

二、西安生态颐养服务中心设计任务书

1. 项目背景

项目用地位于西安市世界地质公园秦岭山脉终南山下子午镇台沟村，长安区生态田园养老社区内，北纬34°02′，东经108°53′，地处陕西省关中平原中部、秦岭北麓、西安市主城区的南部；东临蓝田县，南接宁陕县、柞水县，西与户县、咸阳市接壤，北和雁塔区、灞桥区为邻，距西安市中心15分钟的车程。

本项目定位为养老设施，为西安及周边地区的健康和轻度失能老年人提供长期养老养生服务。竞赛题目结合生态田园养老社区的建设需求，充分应用太阳能等可再生能源技术，利用周边优越的自然环境，建设适用于寒冷地区的绿色、低碳、健康的生态颐养服务中心。

2. 自然条件

项目用地植被茂密，地表景观优美，南临秦岭终南山，空气清新，具有丰富的温泉资源。秦岭是我国南北分界线，号称中国绿肺，蕴藏着金丝猴、朱鹮、大熊猫等丰富的珍稀动植物资源，2009年秦岭终南山被联合国教科文组织评选为世界地质公园。优越的自然地理条件，使秦岭终南山自古以来就成为修道圣地，隐逸佳所，具有悠久的历史文化内涵。

3. 基础设施

基地内基础设施完备，已建有市政自来水、排水、雨水、天然气、供电及通信系统。

Annex1:
Project of Ecological Elderly Care Service Center in Xi'an City

Climate Condition for the Project of Ecological Elderly Care Service Center in Xi'an City

Project site is located in Xi'an world geological park at Tai Gou village Ziwu town at the foot of Zhongnan Mountain of Qinling Mountains, at 34°02′N, 108°53′E in the Wei River Basin in the central Shanxi, and the minimum altitude is 384.7m while the maximum altitude is 2886.7m. The terrain is high in the south and low in the north, high in the east and low in the west, and the south is Qinling Mountains; North is alluvial plain (Daigahara included) of Wei River rift valley; West is Wei River alluvial plain (alluvial fan on northern slope of Qinling Mountains); East is loess tableland and gully of river's channel.

Chang'an district, Xi'an City is the semi humid continental monsoon climate of warm temperate belt, moderate rainfall, mild climate, with distinctive four seasons, short autumn and long spring. Winter is cold and dry; Spring is warm; Summer is hot and rainy; Autumn is mild and humid. The average annual temperature is 13.3°C ; average annual precipitation is 654mm; Humidity is 69.6%; Insolation duration is 1377h. And the temperature for the coldest month, January, is 0.3°C on average and the hottest month, July, is 26.9°C on average. The northeast wind prevails in the region and the annual average wind speed is 1.3 to 2.6m/s. Chang'an is rich in geothermal resources, and geothermal area is about 207.5km², so the energy reservation is equal to the thermal energy produced by 13.254 million tons'coal.

Basic Info
Weather Parameter: 34°18′N, 108°55′E; observation point at 398m.

Month	Temperature °C	Relative Humidity %	Level Solar Radiation kW·h/(m²·d)	Air Pressure kPa	Wind Speed m/s	Ground Temperature °C	Gross Heating Degree °C·d	Gross Cooling Degree °C·d
JAN	0.3	61.4	2.2	89.8	1.5	-4.62	549	0
FEB	3.7	57.9	2.64	89.7	2.0	-0.82	400	0
MAR	8.6	62.3	3.22	89.4	2.2	5.0	291	0
APR	15.3	64.4	4.34	89.1	2.2	13.2	81	159
MAY	20.3	66.1	4.04	88.9	2.3	18.9	0	319
JUN	25.1	61.5	4.5	88.5	2.2	22.8	0	453
JUL	26.9	69.7	4.16	88.4	2.3	24.2	0	524
AUG	25.1	76.4	4.85	88.7	2.3	22.5	0	468
SEP	20.4	77.0	3.08	89.3	1.9	17.8	0	312
OCT	14.3	76.3	3.17	89.7	1.5	11.1	115	133
NOV	7.2	71.5	2.09	89.9	1.5	4.1	324	0
DEC	1.4	65.6	1.75	90	1.4	-2.2	515	0
YEAR	14.1	67.6	3.34	89.3	1.9	11.0	2275	2369

图1　项目所在区位图
Fig. 1　The project's location

图2　项目所在地周边环境
Fig. 2　The project's surroundings

图3　项目用地地形及地表景观
Fig. 3　Terrain and landscape of the project

4. 竞赛场地

本项目所处的生态田园养老社区是由老年养生度假区、老年康复医疗区、老年住宅区、生态颐养服务区以及生态农业园区等共同组成生态田园型综合养老社区。生态颐养服务中心用地位于园区东部、主入口处，建筑红线内分为南高北低两块台地，台地间有2m的高差。

Ecological Elderly Care Service Center in Xi'an City Task Assignment

1. Background

The project is located in Xi'an City, meridian town ditch village, which lies in World Geological Park, Qinling Zhongnan Mountain. It is located in Chang'an District ecological rural nursing community, 34°02′N, 108°53′E. The project is Located in the central Guanzhong Plain of Shanxi province, the north of Qinling Mountains, the south of the main city of Xi'an city. It's on the east of Lantian County, on the south of Ningshan County and Zhashui County, on the west of Huxian Country and Xianyang, and on the north of Yanta District and Baqiao District. It takes 15 minutes' drive to get Xi'an'downtown.

The positioning of the project is elderly facilities. It provides long-term care and health services for people's health and the old with mild disability in Xi'an and surrounding areas. The title of the contest considers the construction demand of ecological rural care community, making full use of solar energy and other renewable energy technologies and surrounding superior natural environment to build a green, low carbon, and healthy ecological maintenance service center that is suitable for the cold area.

2. Natural Condition

The project is covered with plants and the landscape is beautiful. It's on the south of Qinling Zhongnan Mountains, so its air is fresh and it has rich hot spring resources. Qinling Mountains is the boundary between the north and south of China, and known as the China green lung, where there are many resources such as golden monkey, crested ibis, giant pandas and other rare animals and plants. In 2009, Qinling Zhongnan Mountains is chosen as world geology park by UNESCO. Its superior natural and geographical conditions make Qinling Zhongnan Mountains become a religious shrine since ancient times and it's good for seclusion with a long history and owns cultural connotation.

3. Infrastructure

The infrastructure is complete, and it has built municipal water supply, drainage, rain water, natural gas, electricity power supply and communications system.

4. Competition Site

Ecological rural nursing community where the project lies in is an ecological rural comprehensive nursing community which is composed by the elderly health resort, geriatric rehabilitation medical district, old residential areas, ecological cozy service area and ecological agriculture park. Cozy service center of ecological land is located in the east of the park, the main entrance. The building red line is divided into two platforms which is high in the south and low in the north. The relative height between the two platforms is 2m.

5. Design Requirement

(1) Design a elderly care Service Center in the given competition site. Building area is 3200m^2 （±5%）.

(2) The community should has outdoor activity site (the area not less than 200m^2). It also should have a sunshine hall that can control temperature and humidity (not included in the building area and not less than 200m^2). The site should lie in a sunny, sheltered place. The site should ensure that the area of 1/2 is outside the shadow of the local standard buildings.

5. 设计要求

(1) 在给定的竞赛用地范围内设计一处颐养服务中心,建筑面积为3200m²(±5%)。

(2) 用地内应设有室外活动场地(用地不小于200m²),另宜设置可进行温湿度控制调节的阳光活动厅(不计入建筑面积,不小于200m²)。活动场地位置宜选择在向阳、避风处,场地内应保证有1/2的面积处于当地标准的建筑日照阴影之外。

(3) 建筑层数不宜超过3层,并应设置无障碍电梯,且至少一台为医用电梯。

(4) 养老设施建筑及其场地均应进行无障碍设计。

(5) 老年人居住用房和主要公共活动用房应布置在日照充足、通风良好的地段,居住用房冬至日满窗日照不宜小于2h。

(6) 总平面内的道路宜实行人车分流,除满足消防、疏散、运输等要求外,还应保证救护车辆通畅到达所需停靠的建筑物出入口。

(7) 总平面内应设置机动车(不少于15辆)和非机动车停车场。

(8) 养老设施建筑供老年人使用的出入口不应少于两个,其出入口至机动车道路之间应留有缓冲空间。

(9) 建筑用房设置如下表所示:

建筑用房设置　　　　　　　表1

功能			参考面积（含交通面积）	备注
居住生活用房	单人房		总床位80床,房间数量可自行确定	带室内卫生间（含洗浴）
	双人房		1600m²（含交通面积）	
	双人套房			
生活用房	生活辅助用房	公用卫生间 □	350m²（含交通面积）	
		公用厨房 △		
		公共餐厅 □		可兼活动室,并附设备,座位数按总床位数的70%测算
		自助洗衣间 △		
		开水间 □		
		护理站 □		附设护理员值班室、储藏间,并设独立卫浴
		污物间 □		
		交往空间 □		
	生活服务用房	老年人专用浴室 □		附设厕卫
		理发室 □		
		商店 △		

(3) The building should not be more than 3 layers and should have barrier-free elevator, with at least one for the medical elevator.

(4) The construction of elderly facilities and site should consider barrier-free design.

(5) Living housing and public housing for the elderly should be arranged in a sunny, well ventilated area. The living room's daylight shouldn't less than 2 hours on the winter solstice.

(6) Roads in the general plane should be diverted to pedestrians and vehicles. In addition to meeting the requirements of fire control, evacuation, transportation and other requirements, it also should ensure that the ambulance can reach the entrance of the building easily.

(7) In the community, it should have motor vehicles (not less than 15) and non-motor vehicle parking lot.

(8) The entrance in ecological rural care community shouldn't be less than 2. There should be buffer space between the entrance and motor vehicle lane

(9) The building is designed as following table:

the setting of the building　　　Tab. 1

Function			Reference area (including traffic area)	Remarks
Habitable living room	Single room		1600m² (including traffic area)	With indoor bathroom (including bath)
	Double room			
	Double suite			
Living room	Living auxiliary building	Public toilet □	350m² (including traffic area)	
		Public kitchen △		
		Public restaurant □		Also serve as the activity room with equipment, the number of seats is equal to the number beds of 70%
		Self-service laundry room △		
		Boiler Room □		
		Nursing station □		Have nurse duty room, storage room, and an independent bathroom
		Rubbish room □		
		Communication hall □		
	The room for living service	Private bathroom for the elderly people □		A toilet
		Barbershop □		
		Shop △		
Housing for medical care	Medical room	Infirmary □	200m² (including traffic area)	Not less than 2 beds
		Observation room △		
		Disposal room □		
	Housing for health care	Health care room □		
		Psychological counseling room △		

续表

功能				参考面积 (含交通面积)	备注
医疗保健用房	医疗用房	医务室	□	200m² (含交通面积)	不少于两张床
		观察室	△		
		处置室	□		
	保健用房	保健室	△		
		心理疏导室	△		
公共活动用房	活动室	阅览室	△	550m² (含交通面积)	
		网络室	△		
		棋牌室	□		
		书画室	△		
		健身室	□		
	多功能厅		△		
管理服务用房	总值班室		□	500m² (含交通面积)	
	入住登记室		□		
	办公室		□		2~3间
	接待室		□		
	会议室		△		
	档案室		□		
	洗衣房		□		
	职工用房		□		可含职工休息室、职工沐浴间、卫生间、职工餐厅
	备品库		□		
	设备用房		□		

注：① □为应设置；△为宜设置；

② 给出的面积指标可根据设计需求在满足《养老设施建筑设计规范》(GB 50867-2013)的要求下进行调整。

③ 房间在无相互干扰且满足使用功能的前提下可合并设置，多功能使用。

④ 每个养护单元均应设护理站，且位置应明显易找，并宜适当居中。

⑤ 老年人自用卫生间应满足老年人盥洗、便溺的需要，并应留有助厕、助浴等操作空间。

⑥ 公共活动用房应有良好的天然采光与自然通风条件，其位置应避免对老年人卧室产生干扰，平面及空间形式应适合老年人活动需求，并应满足多功能使用的要求。

⑦ 职工用房应考虑工作人员休息、洗浴、更衣、就餐等需求，设置相应的空间。

⑧ 各类用房不应小于下表面积指标：

续表

Function				Reference area (including traffic area)	Remarks
The room for public activity	Activity room	Reading room	△	550m² (including traffic area)	
		Network room	△		
		Chess and card room	□		
		Calligraphy and painting room	△		
		Gymnasium	□		
	Multifunction room		△		
The room for administration	Total duty room		□	500m² (including traffic area)	
	Registration room		□		
	Office		□		2 or 3 beds
	Reception room		□		
	Meeting room		△		
	File room		□		
	Laundry room		□		
	Staff room		□		Have staff rest room, staff bath room, toilet, and staff canteen
	Store room		□		
	Facility room		□		

Note:

① □ stands for "should set up", △ stands for "had better set up".

② The given area index can be adjusted according to the design requirements to meet the requirements of Design Code for Building of Elderly Facilities (50867-2013 GB).

③ The rooms can be combined or multi-function if they don't mutual interference and can meet the use function.

④ Each nursing unit should set up a nursing station.

⑤ Private bathroom for the elderly should meet their washing, defecate need, and should have operation spaces to help the toilet, bath and others.

⑥ Public housing should have good natural lighting and natural ventilation conditions. Its location should avoid interference to the elderly in the bedroom. Planar and spatial forms should be suitable for the activities needs of the elderly and should meet the requirements of multi-functional use.

⑦ Staff housing should consider the staff's rest, bath, changing clothes, dining and other needs, and set the appropriate space.

⑧ Various types of housing should not be less than the following surface area index:

颐养服务中心各类用房最小面积指标　　　表2

用房类别	面积指标	养老院 (m²/床)	备注
老年人用房	生活用房	14.0	不含阳台
	医疗保健用房	2.0	
	公共活动用房	5.0	不含阳光厅/风雨廊
管理服务用房		6.0	

（10）应设集中供暖系统，供暖方式宜选用低温热水地板辐射供暖。

养老设施建筑有关房间的室内冬季供暖计算温度　　　表3

房间	居住用房	生活辅助用房	含沐浴的用房	生活服务用房	活动室多功能厅	医疗保健用房	管理服务用房
计算温度(℃)	20	20	25	18	20	20	18

（11）结合区域建筑特色及气候特点，分析颐养服务中心使用能耗及应用特点，因地制宜地选择和应用主、被动太阳能利用技术及其他可再生能源技术，解决冬季集热采暖和夏季通风降温问题，并考虑技术的经济性，能够实际应用和示范推广。

（12）建筑具备扩建能力和空间改造能力，考虑未来床位增加时的建筑规模扩展和空间功能变更的可能性。

附件2：
泉州生态颐养服务中心项目

一、泉州生态颐养服务中心项目气候条件

项目用地位于世界陶瓷之都——泉州市德化县雷峰镇瓷都印象生态园项目内，北纬25°56′，东经118°32′，项目地处福建省中部，泉州西北部。东与永泰县、莆田市仙游县毗邻，南和永春县接壤，西连大田县，北毗尤溪县，距县城7km，10分钟的车程，西临城关至水口镇355国道，海拔480m以上。

德化县地处闽中屋脊戴云山区，全县地势较高、地形复杂，地貌以山为主，属中亚热带季风气候，平均气温18.2℃，年均无霜期260天左右。境内资源丰富，拥有"山多、水足、矿富、瓷美"四大优势，素有"闽中宝库"之称。山多，全县海拔1000m以上的山峰有258座，福建第二大山脉戴云山主峰横亘境内，是典型的山区县；年均降水量1800mm左右。

All kinds of housing minimum area index in the nursing service　Tab. 2

Area index	Room type	Geracomium (m²/bed)	Remarks
Elderly housing	Living room	14.0	Not include balcony
	Housing for medical care	2.0	
	Housing for public activity	5.0	Not include sun hall/rain gallery
Housing for management		6.0	

(10) Central heating system should be set up. Heating method had better choose low temperature hot water floor radiant heating.

The indoor winter heating temperature of the room in nursing facilities　Tab. 3

Room	Living room	Living auxiliary building	Housing with bath	Housing for living service	Multi hall with activity room	Housing for medical care	Housing for management
Temperature (℃)	20	20	25	18	20	20	18

(11) We combine regional architectural features and climate characteristics, and analyze energy consumption and application characteristics of nursing service center. We choose and apply the active solar energy utilization technology and the passive solar energy utilization technology as well as other renewable energy technology. We should solve the problem of heating in winter and cooling in summer, take in account the economy of technology and be able to practically apply and generalize it.

(12) The building can be extended and transformed considering the possibility of building scale expansion and space function changes in the future if beds are added.

Annex2：
Project of Ecological Elderly Care Service Center in Quanzhou City

Climate Condition for the Project of Ecological Elderly Eare Service Center in Quanzhou City

Project site is located in the world of ceramics—Dehua County of Quanzhou City, Lei Feng Zhen porcelain impression ecological garden project, at 25°56′N, 118°32′E on northwest of Quanzhou, the central part of Fujian Province. East is adjacent to Yongtai county, Xianyou county, Putian city, South borders on Yongchun county, with Datian county in the west and Youxi county in the north, 7km away from the county, for only 10minutes'drive, the west is near

基本气象资料

气象参数：北纬（25°56'）、东经（118°32'）、测量点海拔高度323m。

月份	空气温度 °C	相对湿度 %	水平面日太阳辐射 kW·h/(m²·d)	大气压力 kPa	风速 m/s	土地温度 °C	月采暖度日数 °C·d	供冷度日数 °C·d
一月	8.86	76.3	2.76	98.5	5.25	9.30	276	0
二月	10.4	77.2	2.72	98.3	5.34	11.1	208	0
三月	13.7	78.7	2.88	98.1	5.09	14.7	139	9
四月	18.3	80.6	3.60	97.7	4.58	19.7	39	47
五月	21.9	80.6	3.87	97.4	4.30	23.4	2	123
六月	24.4	83.3	4.30	97.0	4.15	25.6	0	196
七月	25.8	82.2	5.32	97.0	3.99	26.9	0	249
八月	25.4	82.5	4.67	96.9	4.22	26.3	0	234
九月	23.2	81.0	3.87	97.4	4.62	24.0	0	166
十月	19.8	75.1	3.57	97.9	5.38	20.5	8	75
十一月	15.6	73.6	3.07	98.3	5.54	16.1	78	16
十二月	10.7	73.1	2.94	98.5	5.18	11.1	216	1
年平均数	18.2	78.7	3.63	97.7	4.80	19.1	966	1116

二、泉州生态颐养服务中心项目设计任务书

1. 项目背景

项目用地位于世界陶瓷之都——泉州市德化县雷峰镇瓷都印象生态园内，北纬25°56'，东经118°32'，地处福建省中部，泉州西北部；紧邻203省道，距德化县城7km。

瓷都印象生态园是集生态人居、温泉度假、运动休闲、商务会议、星级酒店为一体的养老养生度假小镇。本项目定位于养老设施，为福建及周边地区健康和轻度失能老年人提供长期养老养生服务，并向居住在瓷都印象生态园内的老年人提供社区养老服务。本项目结合德化瓷都印象生态园的定位，充分应用太阳能等可再生能源技术，利用周边优越的自然环境，建设适用于中亚热带气候区的绿色、低碳、健康的生态颐养服务中心。

2. 自然条件

项目用地植被茂密，地表景观优美，空气清新，具有丰富的温泉资源。德化境内山多、水足、矿富、瓷美，素有"闽中宝库"之称。森林覆盖率77.3%，蕴藏着中草药、云豹、黄腹角雉、苏门羚等丰富的珍稀动植物资源。

the national road 355 which connects Chengguan and Shuikou town, and the altitude is above 480m.

Dehua county is located at Daiyun mountain area in central part of Fujia, which has relatively high terrain, complex landform, with a lot of mountains. The county is in the subtropical monsoon climate belt, and the average temperature is 18.2 ℃, and annual average frost-free period lasts around 260 days. There are abundant resources, with "many mountains, plentiful water, rich minerals, delicate china" four advantages, known as the "treasure of central Fujian". There are 258 Mountains which are above 1000m because of the topographic feature of the county, Fujian's second largest mountain Daiyun mountain lie across the county which is typical a mountainous area; the average annual precipitation is about 1800mm.

Basic Info

Weather Parameter: N 25°56', E118°32'; observation point at 323m.

Month	Temperature °C	Relative humidity %	Level Solar Radiation Per Day kW·h/(m²·d)	Air Pressure kPa	Wind Speed m/s	Ground Temperature °C	Monthly Heating Degrees °C·d	Cooling Degree Days °C·d
JAN	8.86	76.3	2.76	98.5	5.25	9.30	276	0
FEB	10.4	77.2	2.72	98.3	5.34	11.1	208	0
MAR	13.7	78.7	2.88	98.1	5.09	14.7	139	9
APR	18.3	80.6	3.60	97.7	4.58	19.7	39	47
MAY	21.9	80.6	3.87	97.4	4.30	23.4	2	123
JUN	24.4	83.3	4.30	97.0	4.15	25.6	0	196
JUL	25.8	82.2	5.32	97.0	3.99	26.9	0	249
AUG	25.4	82.5	4.67	96.9	4.22	26.3	0	234
SEP	23.2	81.0	3.87	97.4	4.62	24.0	0	166
OCT	19.8	75.1	3.57	97.9	5.38	20.5	8	75
NOV	15.6	73.6	3.07	98.3	5.54	16.1	78	16
DEC	10.7	73.1	2.94	98.5	5.18	11.1	216	1
YEAR	18.2	78.7	3.63	97.7	4.80	19.1	966	1116

Ecological Elderly Care Service Center in Quanzhou City Task Assignment

1.Background

The project is located in the world of ceramics—Dehua County of Quanzhou City, Lei Feng Zhen porcelain impression ecological garden project, at 25°56'N, 118°32'E on northwest of Quanzhou, the central part of Fujian Province; it is adjacent to the provincial road 203,7km away from Dehua County.

Capital of ceramic impression ecological park is a town for resort and health cultivation including ecological residential, resort and spa, sports and leisure, business meetings, star hotel. The positioning of the project is elderly facilities which can provide long-term health cultivation service for old people

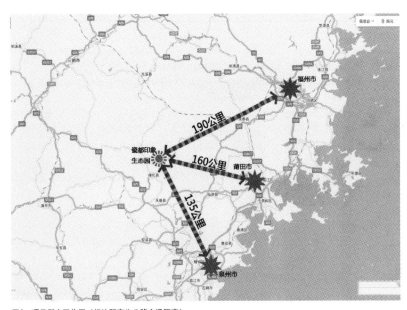

图1 项目所在区位图（标注距离为公路交通距离）
Fig. 1 project location map (marked distance for road transportation)

3. 基础设施

基地内基础设施完备，市政自来水、排水、雨水、天然气、供电及通信系统已同步配套。

4. 竞赛场地

本项目所处的瓷都印象生态园是由养老公寓、养老居住小区、休闲公园、温泉公园、植物园、度假村等共同组成的养老生态园综合项目。生态颐养服务中心用地位于园区西部、主要入口处，交通便利，建筑红线内场地为平地。

5. 设计要求

（1）在给定的竞赛用地范围内设计一处颐养服务中心，建筑面积3200m²（±5%）。

（2）用地内应设有室外活动场地（用地不小于300m²）。

（3）建筑层数不宜超过3层，并应设置无障碍电梯，且至少一台为医用电梯。

（4）养老设施建筑及其场地均应进行无障碍设计。

（5）总平面内的道路宜实行人车分流，除满足消防、疏散、运输等要求外，还应保证救护车辆通畅到达所需停靠的建筑物出入口。

（6）总平面内应设置机动车（不少于5辆）和非机动车停车场。

（7）养老设施建筑供老年人使用的出入口不应少于两个，其出入口至机动车道路之间应留有缓冲空间。

（8）当地地处多雨、多台风地区，建筑宜做遮风避雨措施。

（9）建筑用房设置如下表所示：

who are health or mild disability in Fujian and the surrounding area and provide community service for the old people living in porcelain impression ecological park. This project is combined with the position of ecological park of Dehua porcelain impression, making full use of solar energy and other technologies of renewable energy, utilizing peripheral superior natural environment, to construct green, low carbon and health service center for ecological elderly care, which is suitable to subtropical climate region.

2. Natural Condition

The project land's vegetation is thick; the surface landscape is beautiful; the air is fresh; the hot spring resources are rich. There are abundant mountains, with "many mountains, plentiful water, rich minerals, delicate china", known as the "treasure of central Fujian". The forest coverage rate reaches 77.3%, which contains Chinese herbal medicine, leopard, Tragopan, serow and other rich rare animal and plant resources.

3. Infrastructure

The infrastructure is complete, equipped with municipal water supply, drainage, rain water, natural gas, electricity power supply and communication systems.

4. Competition Site

The project of porcelain impression ecological park is integrated project which is made up with the apartment for the aged people, pension residential estate, leisure park, hot spring park, botanical garden, holiday village and other pension ecological park. The elderly care service center located in the west of ecological park, with convenient transportation in the main entrance, and flat ground within the line of construction site.

5. Design Requirements

(1) The maintenance service center should be designed in the given competition area, and the building area is 3200m² (+5%).

(2) It should provide outdoor playgrounds (no less than 300m²).

(3) The building should not be more than 3 floors, and should set up barrier free elevators, with at least one for the medical elevator.

(4) The construction of elderly facilities and site should consider barrier-free design.

(5) The road in the general plane should be diverted to meet the requirements of fire fighting, evacuation, transportation and so on, and should also ensure that the ambulance can arrive the entrance of the building that it should stop without obstacle.

(6) It should be set up parking lot for motor vehicles (not less than 5) and non-motor vehicle.

(7) The entrance, pension facilities for the elderly to use, should not be less than two, it should be left with a buffer space between the entrance and the road for motor vehicles.

(8) Local place lies in the rainy and typhoon area, so the shelter facilities are suitable for buildings.

(9) The building housing is set as the following table:

建筑用房设置　　　　　　　　表1

功能			参考面积 （含交通面积）	备注
居住生活用房	单人房 双人房 双人套房		1000m²	带室内卫生间 （含洗浴） 总床位50(±2)床房间数量可自行确定
生活用房	生活辅助用房	公用卫生间 □	600m² （含交通面积）	
		公用厨房 △		
		公共餐厅 □		可兼活动室，并附设备需考虑对外开放
		自助洗衣间 △		
		开水间 □		
		护理站 □		附设护理员值班室、储藏间，并设独立卫浴
		污物间 □		
		交往厅 □		
	生活服务用房	老年人专用浴室 □		附设厕卫
		理发室 □		
		商店 △		
		银行、邮电、保险代理 △		
医疗保健用房	医疗用房	医务室 □	250m² （含交通面积）	
		观察室 △		不少于两张床
		治疗室 △		不少于两张床
		药械室 □		
		处置室 □		
		理疗室 □		
	保健用房	保健室 △		
		康复室 △		
		心理疏导室 △		
公共活动用房	活动室	阅览室 △	850m²	
		网络室 △		
		棋牌室 □		
		书画室 △		
		健身室 □		
	多功能厅 △			设置休息区
	阳光厅/风雨廊 △			

the setting of the building　　　　Tab. 1

Function			Reference area (including traffic area)	Remarks
Residence housing	Single room Double room Double suite		1000m²	With indoor bathroom (with bath). The number of beds is 50 (±2). The number of rooms can be determined by themselves
Living housing	Assisted living house	Public toilet □	600m² (Including transportation area)	
		Public kitchen △		
		Public restaurant □		The activity room, and auxiliary equipment need to open up
		Self-service laundry room △		
		Boiler Room □		
		Nursing station □		Add nursing staff duty room, storage room, and a bathroom
		Rubbish room □		
		Communication hall □		
	Life service room	Private bathroom for the elderly people □		Attached toilet
		Barbershop □		
		Shop △		
		Bank, post and telecommunications, insurance agent △		
House for medical care	Medical room	Infirmary □	250m² (Including transportation area)	
		Observation room △		No less than two beds
		Therapeutic room △		No less than two beds
		Medical equipment room □		
		Disposal room □		
		Physiotherapy room □		
	House for health care	Health care room △		
		Recovery room △		
		Psychological counseling room △		

续表

功能			参考面积（含交通面积）	备注
管理服务用房	总值班室	□	500m²	
	入住登记室	□		
	办公室	□		2-3间
	接待室	□		
	会议室	△		
	档案室	□		
	洗衣房	□		
	职工用房	□		可含职工休息室、职工沐浴间、卫生间、职工餐厅
	备品库	□		
	设备用房	□		

注：① □为应设置；△为宜设置；

② 给出的面积指标可根据设计需求在满足《养老设施建筑设计规范》（GB 50867-2013）的要求下进行调整。

③ 房间在无相互干扰且满足使用功能的前提下可以合并设置，多功能使用。

④ 生活用房、公共活动用房应面向园区开放，方便园区内及周边的老年人使用。

⑤ 每个养护单元均应设护理站，且位置应明显易找，并宜适当居中。

⑥ 老年人自用卫生间应满足老年人盥洗、便溺的需要，并应留有助厕、助浴等操作空间。

⑦ 公共活动用房应有良好的天然采光与自然通风条件，其位置应避免对老年人卧室产生干扰，平面及空间形式应适合老年人活动需求，并应满足多功能使用的要求。

⑧ 职工用房应考虑工作人员休息、洗浴、更衣、就餐等需求，设置相应的空间。

续表

Function			Reference area (including traffic area)	Remarks
The room for public activity	Activities room	Reading room △	850m²	
		Internet room △		
		Chess and card room □		
		Painting and calligraphy room △		
		Gymnasium □		
	Multifunction room △			Set rest area
	Sunshine hall/rain gallery △			
The room for administration	Total duty room □		500m²	
	Registration room □			
	Office □			Two or three room
	Reception room □			
	Meeting room △			
	File room □			
	Laundry room □			
	Staff room □			Including lounge, bathroom, toilet, and restaurant for staff
	Store room □			
	Facility room □			

Note:

① □ stands for "should set up", △ stands for "had better set up".

② The area index can be adjusted according to the requirements of design demands to meet the requirements of Design Code for Building of Elderly Facilities (50867-2013 GB).

③ The room can be combined and multifunctional used under the premise of no mutual interference and meeting the use function.

④ The life housing, the room for public activity should be open for the park, so that it is convenient for the elderly who live in and surround the park to use.

⑤ Each unit should set nurse station, and the position should be obvious and easy to find, and it should be properly in the center.

⑥ Toilet which used by old people should meet their need for using the bathroom, urinating, and it should be left operations space to help the toilet, bath and other.

⑦ The room for public activity should be good in natural lighting and ventilation conditions, and its position should avoid interference to old bedroom, and the planar and spatial forms should be suitable for the elderly people's needs, and meet the use need.

⑧ Staff room should consider the needs for staffs' bathing, dressing, dining and so on, setting the corresponding space.

颐养服务中心各类用房最小面积指标　　　表2

用房类别	面积指标	养老院 (m²/床)	备注
老年人用房	生活用房	14.0	不含阳台
	医疗保健用房	2.0	
	公共活动用房	5.0	不含阳光厅/风雨廊
管理服务用房		6.0	

（10）建筑风格宜采用合院式建筑布局，注重院落空间设计。

（11）因地制宜地应用太阳能及其他可再生能源系统，合理选择和应用主、被动太阳能利用技术，着重解决建筑夏季通风降温和空气潮湿、冬季阴冷等问题，创造适老、健康、舒适、生态、环保的绿色空间。

（12）建筑具备空间改造能力，考虑未来使用需求变化而进行空间功能变更的可能性。

The minimum area index of all kinds of housing in maintenance service center　　Tab. 2

Area index	Room type	Geracomium (m²/bed)	Remarks
Elderly housing	Lift housing	14.0	Without balcony
	Health care housing	2.0	
	Public activity housing	5.0	Without sunshine hall/rain gallery
Management service housing		6.0	

(10) He architectural style should adopt the building layout of courtyard style, and pay attention to the courtyard's space design.

(11) Solar energy and other renewable energy systems should be used according to local conditions, with reasonable choice and application of active and passive solar energy technology, mainly solving the problem of the ventilation and cooling heat and humid air in summer, cold in winter and other problems, creating green space which is suitable for the old, healthy, comfortable, ecological and environmental.

(12) The building has the capacity to transform the space and consider the possibility of changing the space function according to the change of use need in the future.

附件3：／Annex 3：
专业术语　Professional Glossary

百叶通风	shutter ventilation
保温	thermal insulation
被动太阳能利用	passive solar energy utilization
敞开系统	open system
除湿系统	dehumidification system
储热器	thermal storage
储水量	water storage capacity
穿堂风	through-draught
窗墙面积比	area ratio of window to wall
次入口	secondary entrance
导热系数	thermal conductivity
低能耗	lower energy consumption
低温热水地板辐射供暖	low temperature hot water floor radiant heating
地板辐射采暖	floor panel heating
地面层	ground layer
额定工作压力	nominal working pressure
防潮层	wetproof layer
防冻	freeze protection
防水层	waterproof layer
分户热计量	household-based heat metering
分离式系统	remote storage system
风速分布	wind speed distribution
封闭系统	closed system
辅助热源	auxiliary thermal source
辅助入口	accessory entrance
隔热层	heat insulating layer
隔热窗户	heat insulation window
跟踪集热器	tracking collector
光伏发电系统	photovoltaic system
光伏幕墙	PV façade
回流系统	drainback system
回收年限	payback time
集热器瞬时效率	instantaneous collector efficiency
集热器阵列	collector array
集中供暖	central heating
间接系统	indirect system
建筑节能率	building energy saving rate
建筑密度	building density
建筑面积	building area
建筑物耗热量指标	index of building heat loss
节能措施	energy saving method
节能量	quantity of energy saving
紧凑式太阳热水器	close-coupled solar water heater
经济分析	economic analysis
卷帘外遮阳系统	roller shutter sun shading system
空气集热器	air collector
空气质量检测	air quality test (AQT)
立体绿化	tridimensional virescence
绿地率	greening rate
毛细管辐射	capillary radiation
木工修理室	repairing room for woodworker
耐用指标	permanent index
能量储存和回收系统	energy storage & heat recovery system
平屋面	plane roof
坡屋面	sloping roof
强制循环系统	forced circulation system
热泵供暖	heat pump heat supply
热量计量装置	heat metering device
热稳定性	thermal stability
热效率曲线	thermal efficiency curve
热压	thermal pressure
人工湿地效应	artificial marsh effect
日照标准	insolation standard
容积率	floor area ratio
三联供	triple co-generation
设计使用年限	design working life
使用面积	usable area
室内舒适度	indoor comfort level
双层幕墙	double façade building

太阳方位角	solar azimuth	自然通风	natural ventilation
太阳房	solar house	自然循环系统	natural circulation system
太阳辐射热	solar radiant heat	自行车棚	bike parking
太阳辐射热吸收系数	absorptance for solar radiation	检查室	examination room
太阳高度角	solar altitude	缴费室	payment
太阳能保证率	solar fraction	康复病房	rehabilitation ward
太阳能带辅助热源系统	solar plus supplementary system	介护病房	involved care ward
太阳能电池	solar cell	护士站	nurse station
太阳能集热器	solar collector	监护室	care unit
太阳能驱动吸附式制冷	solar driven desiccant evaporative cooling	处置室	treatment room
太阳能驱动吸收式制冷	solar driven absorption cooling	针灸室	acupuncture room
太阳能热水器	solar water heating	按摩室	massage room
太阳能烟囱	solar chimney	机械按摩室	mechanical massage room
太阳能预热系统	solar preheat system	调剂室	prescription making up
太阳墙	solar wall	中药配置室	prescription making up for traditional Chinese medicine
填充层	fill up layer		
通风模拟	ventilation simulation	餐厅	canteen
外窗隔热系统	external windows insulation system	厨房	kitchen
温差控制器	differential temperature controller	库房	storehouse
屋顶植被	roof planting	更衣室	locker room
屋面隔热系统	roof insulation system	淋浴	shower bath
相变材料	phase change material (PCM)	运动机能康复室	revcovery room of movement function
相变太阳能系统	phase change solar system	垫上运动区	exercise area on the mat
相变蓄热	phase change thermal storage	听力/语言治疗康复室	treatment and recovery room for hearing and speaking
蓄热特性	thermal storage characteristic		
雨水收集	rain water collection	作业治疗康复室	occupational therapy (OT) room
运动场地	schoolyard	图书室	library
遮阳系数	sunshading coefficient	游艺棋牌室	room for recreation, chess and cards
直接系统	direct system	吸烟室	smoking room
值班室	duty room	电子娱乐室	electronic emtertainment room
智能建筑控制系统	building intelligent control system	办公室	office
中庭采光	atrium lighting	总务室	general office
主入口	main entrance	会议室	meeting room
贮热水箱	heat storage tank	档案室	file room
准备室	preparation room	医务科	medical services
准稳态	quasi-steady state		